NUMBER THEORY

NUMBER THEORY

S.K. Jain

IVY PUBLISHING HOUSE
DELHI - 110095

Published by
IVY PUBLISHING HOUSE
Delhi-110095

Sole Distributors
SARUP & SONS
4740/23, Ansari Road
Darya Ganj, New Delhi-110002
Tel.: 23281029, 23244664, 41010989
Fax: 011-23277098
E-mail: sarupandsonsin@hotmail.com

NUMBER THEORY

© Reserved

Edition : 2022

ISBN 81-7890-174-9

PRINTED IN INDIA

Laser Typesetting at Twinkle Graphics and Printed at Roshan Offset Printers, Delhi.

Preface

This book gives an undergraduate-level introduction to Number Theory, with the emphasis on fully explained proofs and examples; exercises (with solutions) are integrated into the text. The first few chapters, covering divisibility, prime numbers and modular arithmetic, assume only basic school algebra, and are therefore suitable for first or second year students as an introduction to the methods of pure mathematics. Elementary ideas about groups and rings (summarised in an appendix) are then used to study groups of units, quadratic residues and arithmetic functions with applications to enumeration and cryptography. The final part, uses ideas from algebra, analysis, calculus and geometry to study Dirichlet series and sums of squares; in particular, the last chapter gives a concise account of Fermat's Last Theorem, from its origin in the ancient Babylonian and Greek study of Pythagorean triples to its recent proof by Andrew Wiles.

S.K. Jain

Contents

1. Number Theory — 1
2. Identities for Linear Recurring Sequences — 26
3. Half-Totient Tree — 28
4. Limit Cycles of xy (mod x+y) — 34
5. Rounding Up To PI — 41
6. Fermat's Last Theorem for Cubes — 46
7. Digit Reversal Sums Leading to Palindromes — 55
8. Discordance Impedes Square Magic — 61
9. Least Significant Non-Zero Digit of n! — 70
10. Geodesic Diophantine Boxes — 76
11. Highly Heronian Ellipses — 87
12. How Leibniz Might Have Anticipated Euler — 96
13. Odd-Greedy Unit Fraction Expansions — 102
14. Four Squares from Three Numbers — 111
15. Accidental Melodies — 117
16. Some Properties of the Lucas Sequence — 125
17. On a Unit Fraction Question of Erdos and Graham — 137
18. The Greedy Algorithm for Unit Fractions — 139
19. Average of Sigma(n)/n — 147
20. Lucas's Primality Test With Factored N-1 — 152
21. One In The Chamber — 155
22. Fractions and Characteristic Recurrences — 157
23. Automedian Triangles and Magic Squares — 160
24. Orthomagic Square of Squares — 168
25. Magic Square of Squares — 174
25. Magic Square of Squares — 174

26.	Anti-Carmichael Pairs	182
27.	Coherent Arrays of Squares	184
28.	Mock-Rational Numbers	187
29.	Integer Sequences Related To PI	196
30.	Series Within Parallel Resistance Networks	202
31.	Pythagorean Graphs	208
32.	On General Palindromic Numbers	214
33.	Minimizing the Denominators of Unit Fraction Expansions	219
34.	Perrin's Sequence	225
35.	Unit Fraction Partitions	230
36.	Reflective and Cyclic Sets of Primes	236
37.	Waring's Problem	240
38.	Cyclic Divisibility	242
39.	Unit Fractions and Fibonacci	247
40.	Solving Magic Squares	250
41.	Concordant Forms	256
42.	Numbers Expressible As $(a^2 - 1)(b^2 - 1)$	262
43.	Euclidean Algorithm	269
44.	On the Density of Some Exceptional Primes	272
45.	Recurrences and Pell Equations	287

1
Number Theory

Introduction

Number theory is one of the oldest branches of pure mathematics, and one of the largest. Of course, it concerns questions about numbers, usually meaning whole numbers or rational numbers (fractions).

Elementary number theory involves divisibility among integers — the division "algorithm", the Euclidean algorithm (and thus the existence of greatest common divisors), elementary properties of primes (the unique factorization theorem, the infinitude of primes), congruences (and the structure of the sets Z/nZ as commutative rings), including Fermat's little theorem and Euler's theorem extending it. But the term "elementary" is usually used in this setting only to mean that no advanced tools from other areas are used — *not* that the results themselves are simple. Indeed, a course in "elementary" number theory usually includes classic and elegant results such as Quadratic Reciprocity; counting results using the Möbius Inversion Formula (and other multiplicative number-theoretic functions); and even the Prime Number Theorem, asserting the approximate density of primes among the integers, which has difficult but "elementary" proofs. Other topics in elementary number theory — the solutions of sets of linear congruence equations (the Chinese Remainder Theorem), or solutions of single binary quadratic equations (Pell's equations and continued fractions), or the generation of Fibonacci numbers or Pythagorean triples — turn out in retrospect to be harbingers of sophisticated tools and themes in other areas.

The remaining parts of number theory are more or less closely allied with other branches of mathematics, and typically use tools from those areas.

For example, many questions in number theory may be posed as Diophantine equations — equations to be solved in integers — without much preparation. Catalan's conjecture — are 8 and 9 the only consecutive powers? — asks for the solution to $x^a - y^b = 1$ in integers; the Four Squares Theorem — every natural number is the sum of four integer squares — simply asserts that $x^2 + y^2 + z^2 + w^2 = n$ is solvable for all n. But the attempt to solve these equations requires rather powerful tools from elsewhere in mathematics to shed light on the the structure of the problem. (Even the *possibility* of analyzing Diophantine equations — Hilbert's tenth problem — suggests the use of mathematical logic; Matijasevic's negative solution of that problem guarantees number theorists will never find a complete solution to their analyses!)

We can try to subdivide number theory according to those other tools used. Naturally there is significant overlap, and a single question from elementary number theory often requires tools from many branches of number theory.

"Combinatorial Number Theory" involves the number-theoretic study of objects which arise naturally from counting or iteration. This includes a study of many specific families of numbers — the binomial coefficients, the Fibonacci numbers, Bernoulli numbers, factorials, perfect squares, partition numbers and so on — which can be obtained by simple recurrence relations, say, or as values of polynomials. One asks for their factorizations, their congruence properties, their densities, etc. It is very easy to state conjectures in this area which can often be understood without any particular mathematical training, but which can be very difficult to solve; Erdös has left many conjectures of this sort.

"Algebraic Number Theory" extends the concept of "number" to mean an element of some ring, usually the ring of integers in a finite algebraic extension of the rational number field. These arise naturally even when considering elementary topics (e.g. the representation of an integer as a sum of two squares is tantamount to its factorization in the ring $Z[i]$ of Gaussian integers) but are also interesting in their own right. In this setting, the familiar features of the natural numbers (e.g. unique factorization) need not hold. The virtue of the machinery introduced — class groups, discriminants, Galois theory, field cohomology, class field theory, group representations and L-functions — is that it allows a reconstruction of some of that order in these new settings.

A key feature of some problems in number theory is the extent to which the behaviour of the problem in integers is reflected in its behaviour modulo p for all primes p, and its behaviour in the real line. The correct construction for the investigation of this phenomenon is usually a local ring such as the p-adic integers. These fields provide an opportunity for unusual forms of analysis (e.g. series converge iff their terms converge to zero — the calculus student's dream!) Local analysis usually arises as a part of algebraic number theory.

"Analytic Number Theory" involves the study of the Riemann zeta function and other similar functions such as Dirichlet series. The zeta function may be defined on half the complex plane as the sum $1 + 1/2^s + 1/3^s + 1/4^s + \ldots$; its connection with number theory results from its factorization as a product $\operatorname{Prod}(1 - 1/p^s)^{(-1)}$, the product taken over all primes p. Thus for example the distribution of the primes among the integers can be deduced from a good understanding of the behaviour of zeta(s). The Riemann Hypothesis states that zeta(s) is never zero except along the line $\operatorname{Re}(s)=1/2$ (or at the negative even integers). This is arguably the most important open question in mathematics. There are other related functions, useful either for studying the Riemann zeta function or for making similar conclusions about other sets; for example, one may use them to prove the infinitude of primes in candidate linear progressions.

Other areas of number theory are also quite analytical. For example, "additive number theory" asks about ways of expressing an integer N as a sum of integers a_i in a set A. If we set $f(z) = \operatorname{Sum} \exp(2 \pi i\, a_i\, z)$, then $f(z)^k$ has $\exp(2 \pi i\, N\, z)$ as a summand iff N is a sum of k of the a_i. This in turn can be deduced from some integration (the integral of $\exp(a z)$ around a circle is 0 or not depending on whether a is zero). Thus analytical techniques are used to approach Waring's problem, for example (representing integers as sums of squares, cubes, etc.), and to address other questions with exponential sums. Since these computations are equivalent to work in the rings Z/nZ, there is an interest in the structure of these rings.

One may include in this analytic category the parts of number theory connected with forms (e.g. quadratic forms are quadratic polynomials in several variables). Broadly speaking the goal here is to analyze the possible equivalence classes of functions under groups of symmetries. Even when few analytic tools are used for the analysis of the functions themselves, the groups of interest (e.g. the discontinuous groups acting on the complex upper half-plane) are well understood in areas of analysis.

Also, ideas from analysis (measure theory, dimension) can be used in "probabilistic number theory", in which one studies almost-periodic, pseudo-random, or fractal behaviour of number-theoretic functions.

Finally, a significant amount of analysis is also used in Sieve methods, and other aspects of multiplicative number theory. Here one generalizes the sieve of Eratosthenes to investigate the presence of, say, prime pairs (Brun's sieve) or solutions to the Goldbach conjecture (every even number is a sum of two primes).

"Transcendental number theory" considers proofs of transcendence or algebraicity of numbers, and the extent to which numbers can be approximated by algebraic numbers (say). This has a direct bearing on other fields such as Diophantine equations, for example, since the unsolvability of a Diophantine equation can be deduced from the observation that it would require rational numbers which approximate a real number "too well". Well-known results in this area include the transcendence of pi, which in turn shows the impossibility of squaring the circle.

"Geometric number theory" incorporates all forms of geometry. The classical Geometry of Numbers due to Minkowski begins with statements of Euclidean geometry on lattices (A convex body contains a lattice poir t if its volume is large enough); by extension this becomes the study of quadratic forms on lattices, and thus a method of investigating regular packings of spheres, say. But one may also investigate algebraic geometry with number theory, that is, one may study varieties such as algebraic curves and surfaces and ask if they have *rational* or *integral* solutions (points with rational or integral coordinates). This topic includes

Number Theory

the highly successful theory of elliptic curves (where the rational points form a finitely generated group) and finiteness results (e.g. Siegel's, Thue's, or Faltings's) which apply to integral or higher-genus situations.

"Computational number theory" studies the effectiveness of algorithms for computation of number-theoretic quantities. Considerable effort has been expended in primality-testing and integer factorization routines, for example — procedures which are in principle trivial, but whose naive solution is untenable in large cases. This field also considers integer quantities (e.g the class number) whose usual definition is nonconstructive, and real quantities (e.g. the values of zeta functions) which must be computed with very high precision; thus this overlaps both computer algebra and numerical analysis.

History

Fairly comprehensive accounts may be found in
- Weil, André: "Number theory, An approach through history", Birkhäuser Boston, Inc., Boston, Mass., 1984 ISBN 0-8176-3141-0
- Ore, Oystein, "Number theory and its history", Dover Publications, Inc., New York, 1988. 370 pp. ISBN 0-486-65620-9
- Dickson, Leonard Eugene: "History of the theory of Numbers", Carnegie Institution of Washington publication. no. 256, 1919: a three-volume history and literature review through early 20th century

Note that there are many popular books addressing particular aspects of number theory likely to be of wide interest. Some of these are mentioned on index pages for particular subdisciplines. Also likely to be helpful are biographies of mathematicians who have been prominent across much of number theory (especially recent romantic figures such as Ramanujan and Erdös).

Applications and related fields

Clearly the separate parts of number theory overlap with other areas of mathematics.

"Combinatorial number theory", of course, overlaps quite a bit with 05: Combinatorics. While here we consider number-theoretic topics involving the binomial coefficients, partitions, and so on, in combinatorics one might be interested in these numbers as real-number sequences, or as tools for counting sets of some type.

Questions in algebraic number theory often require tools of Galois theory; that material is mostly a part of 12: Field theory (particularly the subject of field extensions).

The algebraic structure of the ring of integers is similar to that of other commutative rings such as rings of polynomials. (Most material on polynomials is on the Galois Theory page.)

The sets of solutions in rational numbers to algebraic equations may be viewed as algebraic varieties, and thus studied with tools of 14: Algebraic Geometry. This is particularly true with single equations in two variables (which lead to curves); such equations when of degree 3 (or 4) lead to elliptic curves.

The theory of lattices is arguably one of quadratic forms, which are considered in more detail within Linear Algebra. (Note that these are essentially unrelated to the lattices studied along with Ordered Sets.)

Sequences and series of integers may be viewed as series and sequences of real numbers, and hence studied with tools of analysis.

Lattices can be used to determine patterns for sphere-packing; consult the sphere FAQ.

There is current interest in factoring large integers (and primality testing), in part because of relationships with 94: cryptography.

The discipline of computing numerical answers to problems (e.g. locating the roots of a polynomial) is not number theory at all but Numerical Analysis.

Other fields with some overlap as seen in the diagram are areas 20 (Group Theory), 81 (Quantum Theory), 22 (Topological groups), 58 (Global Analysis), 01 (History), 68 (Computer Science), 52 (Convex Geometry), 65 (Numerical Analysis), 03 (Logic), 33 (Special Functions). The diagram manages to show that some parts of number theory are related to fields of algebra, while others involve more discrete mathematics or analysis.

Number theory is the branch of pure mathematics concerned with the properties of numbers in general, and integers in particular, as well as the wider classes of problems that arise

Number Theory

from their study.

Number theory may be subdivided into several fields, according to the methods used and the type of questions investigated.

The term "arithmetic" is also used to refer to number theory. This is a somewhat older term, which is no longer as popular as it once was. Number theory used to be called the higher arithmetic, but this too is dropping out of use. Nevertheless, it still shows up in the names of mathematical fields (arithmetic functions, arithmetic of elliptic curves, fundamental theorem of arithmetic). This sense of the term arithmetic should not be confused either with elementary arithmetic, or with the branch of logic which studies Peano arithmetic as a formal system. Mathematicians working in the field of number theory are called **number theorists**.

Elementary number theory

In **elementary number theory**, integers are studied without use of techniques from other mathematical fields. Questions of divisibility, use of the Euclidean algorithm to compute greatest common divisors, factorization of integers into prime numbers, investigation of perfect numbers and congruences belong here. Several important discoveries of this field are Fermat's little theorem, Euler's theorem, the Chinese remainder theorem and the law of quadratic reciprocity. The properties of multiplicative functions such as the Möbius function, Euler's ö function, integer sequences, factorials and Fibonacci numbers all also fall into this area.

Many questions in number theory can be stated in elementary number theoretic terms, but they may require very deep consideration and new approaches outside the realm of elementary number theory. Examples include:
- The Goldbach conjecture concerning the expression of even numbers as sums of two primes.
- Catalan's conjecture (now Mihăilescu's theorem) regarding successive integer powers.
- The twin prime conjecture about the infinitude of prime pairs.

- The Collatz conjecture concerning a simple iteration.
- Fermat's last theorem (stated in 1637, but not proved until 1994) concerning the impossibility of finding nonzero integers x, y, z such that $x^n + y^n = z^n$ for some integer n greater than 2.

The theory of Diophantine equations has even been shown to be *undecidable* (see Hilbert's tenth problem).

Analytic number theory

Analytic number theory employs the machinery of calculus and complex analysis to tackle questions about integers. The prime number theorem and the related Riemann hypothesis are examples. Waring's problem (representing a given integer as a sum of squares, cubes etc.), the Twin Prime Conjecture (finding infinitely many prime pairs with difference 2) and Goldbach's conjecture (writing even integers as sums of two primes) are being attacked with analytical methods as well. Proofs of the transcendence of mathematical constants, such as ð or e, are also classified as analytical number theory. While statements about transcendental numbers may seem to be removed from the study of integers, they really study the possible values of polynomials with integer coefficients evaluated at, say, e; they are also closely linked to the field of Diophantine approximation, where one investigates "how well" a given real number may be approximated by a rational one.

Algebraic number theory

In **algebraic number theory**, the concept of a number is expanded to the algebraic numbers which are roots of polynomials with rational coefficients. These domains contain elements analogous to the integers, the so-called algebraic integers. In this setting, the familiar features of the integers (e.g. unique factorization) need not hold. The virtue of the machinery employed—Galois theory, group cohomology, class field theory, group representations and L-functions—is that it allows to recover that order partly for this new class of numbers.

Many number theoretic questions are best attacked by studying them *modulo p* for all primes p (see finite fields). This is called *localization* and it leads to the construction of the p-

Number Theory

adic numbers; this field of study is called local analysis and it arises from algebraic number theory.

Geometric number theory

Geometric number theory (traditionally called geometry of numbers) incorporates all forms of geometry. It starts with Minkowski's theorem about lattice points in convex sets and investigations of sphere packings.

Combinatorial number theory

Combinatorial number theory deals with number theoretic problems which involve combinatorial ideas in their formulations or solutions. Paul Erdős is the main founder of this branch of number theory. Typical topics include covering system, zero-sum problems, various restricted sumsets, and arithmetic progressions in a set of integers. Algebraic or analytic methods are powerful in this field.

Computational number theory

Computational number theory studies algorithms relevant in number theory. Fast algorithms for prime testing and integer factorization have important applications in cryptography.

Vedic number theory

Mathematicians in India were interested in finding integral solutions of Diophantine equations since the Vedic era. The earliest geometric use of Diophantine equations can be traced back to the Sulba Sutras, which were written between the 8th and 6th centuries BC. Baudhayana (c. 800 BC) found two sets of positive integral solutions to a set of simultaneous Diophantine equations, and also used simultaneous Diophantine equations with up to four unknowns. Apastamba (c. 600 BC) used simultaneous Diophantine equations with up to five unknowns.

Jaina number theory

In India, Jaina mathematicians developed the earliest systematic theory of numbers from the 4th century BC to the 2nd century CE. The Jaina text *Surya Prajinapti* (c. 400 BC) classifies all numbers into three sets: enumerable, innumerable and infinite. Each of these was further subdivided into three orders:

- Enumerable: lowest, intermediate and highest.
- Innumerable: nearly innumerable, truly innumerable and innumerably innumerable.
- Infinite: nearly infinite, truly infinite, infinitely infinite.

The Jains were the first to discard the idea that all infinites were the same or equal. They recognized five different types of infinity: infinite in one and two directions (one dimension), infinite in area (two dimensions), infinite everywhere (three dimensions), and infinite perpetually (infinite number of dimensions).

The highest enumerable number N of the Jains corresponds to the modern concept of aleph-null $\aleph_0$ (the cardinal number of the infinite set of integers 1, 2, ...), the smallest cardinal transfinite number. The Jains also defined a whole system of transfinite cardinal numbers, of which is the smallest.

In the Jaina work on the theory of sets, two basic types of transfinite numbers are distinguished. On both physical and ontological grounds, a distinction was made between *asmkhyata* and *ananata*, between rigidly bounded and loosely bounded infinities.

Hellenistic number theory

Number theory was a favorite study among the Hellenistic mathematicians of Alexandria, Egypt from the 3rd century CE, who were aware of the Diophantine equation concept in numerous special cases. The first Hellenistic mathematician to study these equations was Diophantus.

Diophantus also looked for a method of finding integer solutions to linear indeterminate equations, equations that lack sufficient information to produce a single discrete set of answers. The equation $x + y = 5$ is such an equation. Diophantus discovered that many indeterminate equations can be reduced to a form where a certain category of answers is known even though a specific answer is not.

Number Theory

Classical Indian number theory

Diophantine equations were extensively studied by mathematicians in medieval India, who were the first to systematically investigate methods for the determination of integral solutions of Diophantine equations. Aryabhata (499) gave the first explicit description of the general integral solution of the linear Diophantine equation $ay + bx = c$, which occurs in his text *Aryabhatiya*. This *kuttaka* algorithm is considered to be one of the most significant contributions of Aryabhata in pure mathematics, which found solutions to Diophantine equations by means of continued fractions. The technique was applied by Aryabhata to give integral solutions of simulataneous linear Diophantine equations, a problem with important applications in astronomy. He also found the general solution to the indeterminate linear equation using this method.

Brahmagupta in 628 handled more difficult Diophantine equations. He used the *chakravala* method to solve quadratic Diophantine equations, including forms of Pell's equation, such as $61x^2 + 1 = y^2$. His *Brahma Sphuta Siddhanta* was translated into Arabic in 773 and was subsequently translated into Latin in 1126. The equation $61x^2 + 1 = y^2$ was later posed as a problem in 1657 by the French mathematician Pierre de Fermat. The general solution to this particular form of Pell's equation was found over 70 years later by Leonhard Euler, while the general solution to Pell's equation was found over 100 years later by Joseph Louis Lagrange in 1767. Meanwhile, many centuries ago, the general solution to Pell's equation was recorded by Bhaskara II in 1150, using a modified version of Brahmagupta's *chakravala* method, which he also used to find the general solution to other indeterminate quadratic equations and quadratic Diophantine equations. Bhaskara's *chakravala* method for finding the general solution to Pell's equation was much simpler than the method used by Lagrange over 600 years later. Bhaskara also found solutions to other indeterminate quadratic, cubic, quartic and higher-order polynomial equations. Narayana Pandit further improved on the *chakravala* method and found more general solutions to other indeterminate quadratic and higher-order polynomial equations.

Islamic number theory

From the 9th century, Islamic mathematicians had a keen interest in number theory. The first of these mathematicians was the Arab mathematician Thabit ibn Qurra, who discovered a theorem which allowed pairs of amicable numbers to be found, that is two numbers such that each is the sum of the proper divisors of the other. In the 10th century, Al-Baghdadi looked at a slight variant of Thabit ibn Qurra's theorem.

In the 10th century, al-Haitham seems to have been the first to attempt to classify all even perfect numbers (numbers equal to the sum of their proper divisors) as those of the form $2^{k-1}(2^k - 1)$ where $2^k - 1$ is prime. Al-Haytham is also the first person to state Wilson's theorem, namely that if p is prime then $1 + (p - 1)!$ is divisible by p. It is unclear whether he knew how to prove this result. It is called Wilson's theorem because of a comment made by Edward Waring in 1770 that John Wilson had noticed the result. There is no evidence that John Wilson knew how to prove it and most certainly Waring did not. Lagrange gave the first proof in 1771.

Amicable numbers played a large role in Islamic mathematics. In the 13th century, Persian mathematician Al-Farisi gave a new proof of Thabit ibn Qurra's theorem, introducing important new ideas concerning factorisation and combinatorial methods. He also gave the pair of amicable numbers 17296, 18416 which have been attributed to Euler, but we know that these were known earlier than al-Farisi, perhaps even by Thabit ibn Qurra himself. In the 17th century, Muhammad Baqir Yazdi gave the pair of amicable numbers 9,363,584 and 9,437,056 still many years before Euler's contribution.

Early European number theory

Number theory began in Europe in the 16th and 17th centuries, with François Viète, Bachet de Meziriac, and especially Fermat, whose infinite descent method was the first general proof of diophantine questions. Fermat's last theorem was posed as a problem in 1637, a proof of which wasn't found until 1994. Fermat also posed the equation $61x^2 + 1 = y^2$ as a problem in 1657.

In the eighteenth century, Euler and Lagrange made important

Number Theory

contributions to number theory. Euler did some work on analytic number theory, and found a general solution to the equation $61x^2 + 1 = y^2$, which Fermat posed as a problem. Lagrange found a solution to the more general Pell's equation. Euler and Lagrange solved these Pell equations by means of continued fractions, though this was more difficult than the Indian *chakravala* method.

Beginnings of modern number theory

Around the beginning of the nineteenth century books of Legendre (1798), and Gauss put together the first systematic theories in Europe. Gauss's *Disquisitiones Arithmeticae* (1801) may be said to begin the modern theory of numbers.

The formulation of the theory of congruences starts with Gauss's *Disquisitiones*. He introduced the symbolism

$$a \equiv b \pmod{c},$$

and explored most of the field. Chebyshev published in 1847 a work in Russian on the subject, and in France Serret popularised it.

Besides summarizing previous work, Legendre stated the law of quadratic reciprocity. This law, discovered by induction and enunciated by Euler, was first proved by Legendre in his *Théorie des Nombres* (1798) for special cases. Independently of Euler and Legendre, Gauss discovered the law about 1795, and was the first to give a general proof. To the subject have also contributed: Cauchy; Dirichlet whose *Vorlesungen über Zahlentheorie* is a classic; Jacobi, who introduced the Jacobi symbol; Liouville, Zeller(?), Eisenstein, Kummer, and Kronecker. The theory extends to include cubic and biquadratic reciprocity, (Gauss, Jacobi who first proved the law of cubic reciprocity, and Kummer).

To Gauss is also due the representation of numbers by binary quadratic forms.

Prime number theory

A recurring and productive theme in number theory is the study of the distribution of prime numbers. Carl Friedrich Gauss conjectured the limit of the number of primes not exceeding a given number (the prime number theorem) as a teenager.

Chebyshev (1850) gave useful bounds for the number of

primes between two given limits. Riemann introduced complex analysis into the theory of the Riemann zeta function. This led to a relation between the zeros of the zeta function and the distribution of primes, eventually leading to a proof of prime number theorem independently by Hadamard and de la Vallée Poussin in 1896. However, an elementary proof was given later by Paul Erdős and Atle Selberg in 1949+. Here elementary means that it does not use techniques of complex analysis; however, the proof is still very ingenious and difficult. The Riemann hypothesis, which would give much more accurate information, is still an open question.

Nineteenth-century developments

Cauchy, Poinsot (1845), Lebesgue(?) (1859, 1868), and notably Hermite have added to the subject. In the theory of ternary forms Eisenstein has been a leader, and to him and H. J. S. Smith is also due a noteworthy advance in the theory of forms in general. Smith gave a complete classification of ternary quadratic forms, and extended Gauss's researches concerning real quadratic forms to complex forms. The investigations concerning the representation of numbers by the sum of 4, 5, 6, 7, 8 squares were advanced by Eisenstein and the theory was completed by Smith.

Dirichlet was the first to lecture upon the subject in a German university. Among his contributions is the extension of Fermat's last theorem:

$$x^n + y^n \neq z^n, (x, y, z \neq 0, n > 2)$$

which Euler and Legendre had proven for $n = 3,4$ (and therefore by implication, all multiples of 3 and 4), Dirichlet showing that

$$x^5 + y^5 \neq az^5.$$ Among the later French writers are Borel; Poincaré, whose memoirs are numerous and valuable; Tannery, and Stieltjes. Among the leading contributors in Germany were Kronecker, Kummer Schering, Bachmann, and Dedekind. In Austria Stolz's *Vorlesungen über allgemeine Arithmetik* (1885-86), and in England Mathews' Theory of Numbers (Part I, 1892) were scholarly of general works.

Twentieth-century developments

Major figures in twentieth-century number theory include

Paul Erdős, Gerd Faltings, G. H. Hardy, Edmund Landau, John Edensor Littlewood, Srinivasa Ramanujan and André Weil. Milestones in twentieth-century number theory include the proof of Fermat's Last Theorem by Andrew Wiles in 1994 and the proof of the related Taniyama–Shimura theorem in 1999.

Number theory

The study of the properties and relations of the integers. There are many sets of positive integers of particular interest, such as the primes and the perfect numbers. Number theory, of ancient and continuing interest for its intrinsic beauty, also plays a crucial role in computer science, particularly in the area of cryptography.

Elementary number theory

This part of number theory does not rely on advanced mathematics, such as complex analysis and ring theory. The basic notion of elementary number theory is divisibility. An integer d is a divisor of n, written $d \mid n$, if there is an integer t such that $n = dt$. A prime number is a positive integer that has exactly two positive divisors, 1 and itself. The ten smallest primes are 2, 3, 5, 7, 11, 13, 17, 19, 23, and 29. Euclid (around 300 B.C.) proved that there are infinitely many primes by showing that if the only primes were 2, 3, 5, p, a prime not in this list could be found by taking a prime factor of the number shown in Eq. (1).

$$N = (2 \cdot 3 \cdot 5 \cdots p) + 1 \tag{1}$$

Primes are the building blocks of the positive integers. The fundamental theorem of arithmetic, established by K. F. Gauss in 1801, states that every positive integer can be written as the product of prime factors in exactly one way when the order of the primes is disregarded. The fundamental ingredient in the proof is a lemma proved by Euclid: if a is a divisor of bc and a and b have no common factors, then a divides c. Related web site: Prime Numbers Related web site: The Prime Pages

The greatest common divisor of the positive integers a and b, written $\gcd(a, b)$, is the largest integer that divides both a and *b*. Two integers with greatest common divisor equal to 1 are called coprime. Euclid devised a method, now known as the euclidean algorithm, for finding the greatest common divisor of two positive integers. This method works by replacing a and b, where a b, by b and the remainder when a is divided by b. This step is repeated until a remainder of 0 is reached.

Perfect numbers and Mersenne primes

A perfect number is a positive integer equal to the sum of its positive divisors other than itself. L. Euler showed that $2^{n-1}(2^n - 1)$ is perfect if and only if $2^n - 1$ is prime. If $2^n - 1$ is prime, then n itself must be prime. Primes of the form $2^p - 1$ are known as Mersenne primes after M. Mersenne, who studied them in the seventeenth century. As of 2000, there were 38 Mersenne primes, the largest being $2^{6,972,593} - 1$, a number with 2,098,960 decimal digits. For most of the time since 1876, the largest known prime has been a Mersenne prime; this is the case because there is a special method, known as the Lucas-Lehmer test, to determine whether $2^p - 1$ is prime. Only 12 Mersenne primes were known prior to the invention of computers. Since the early 1950s, with the help of computers, Mersenne primes have been found at a rate averaging approximately one every 2 years. The four largest Mersenne primes were found as part of the Great Internet Mersenne Prime Search (GIMPS). GIMPS provides optimized software for running the Lucas-Lehmer test. Thousands of people participate in GIMPS, running a total of almost 1 teraflops in the search for Mersenne primes. Related web site: Great Internet Mersenne Prime Search (GIMPS)

All even perfect numbers are given by Euler's formula. Whether there are odd perfect numbers is still an unsolved problem. If an odd perfect number existed, it would have at least eight distinct prime factors and be larger than 10^{300}.

Congruences

If $a - b$ is divisible by m, then a is called congruent to b modulo m, and this relation is written a b (mod m). This relation between integers is an equivalence relation and defines equivalence classes of numbers congruent to each other, called residue classes. Congruences to the same modulus can be added, subtracted, and multiplied in the same manner as equations. However, when both sides of a congruence are divided by the same integer d, the modulus m must be divided by $\gcd(d, m)$. There are m residue classes modulo m. The number of classes containing only numbers coprime to m is denoted by (m), where (m) is called the Euler phi function. The value of (m) is given by Eq. (2).

$$\phi(m) = m \prod_{p|m} \left(1 - \frac{1}{p}\right) \tag{2}$$

Fermat's little theorem states that if p is prime and a is coprime to p, then formula (3)

$$a^{p-1} \equiv 1 \pmod{p} \tag{3}$$

is true. Euler generalized this congruence by showing that formula (4)

$$a^{\phi(m)} \equiv 1 \pmod{m} \tag{4}$$

is valid whenever a and m are coprime.

The linear congruence ax b (mod m) is solvable for x if and only if $d \mid b$ where $d = \gcd(a, m)$. Under this condition, it has exactly d incongruent solutions modulo m. If p is prime, then the congruence ax 1 (mod p) has exactly one solution for each a not divisible by p. This solution is the inverse of a modulo p. This implies that the residue classes modulo p form a finite field of p elements. See also: Field theory (mathematics)

The simultaneous system of congruences x a_i (mod m_i), $i = 1, 2, , r$, where the moduli m_i, $i = 1, 2, 3, , r$, are pairwise coprime positive integers, has a unique solution modulo M, where M is the product of these moduli. This result, called the Chinese remainder theorem, was known by ancient Chinese and Hindu mathematicians.

Primality testing and factorization

Since every composite integer has a prime factor not exceeding its square root, to determine whether n is prime, it is necessary only to show that no prime between 2 and n divides n. However, this simple test is extremely inefficient when n is large. Since generating large primes is important for cryptographic applications, better methods are needed.

By Fermat's little theorem, if 2^n is not congruent to 2 modulo

n, then n is not prime; this was known to the ancient Chinese, who may have thought that the converse was true. However, it is possible that $2^n \equiv 2 \pmod{n}$ without n being prime; such integers n are called pseudoprimes to the base 2. More generally, an integer n is called a pseudoprime to the base b if n is not prime but $b^n \equiv b \pmod{n}$. A Carmichael number is a number n that is not prime but that is a pseudoprime to all bases b, where b is coprime to n and less than n. The smallest Carmichael number is 561. In 1992 it was shown that there are infinitely many Carmichael numbers.

A positive integer $n = 2^s t$, where t is odd, is called a strong pseudoprime to the base b if either $b^t \equiv 1 \pmod{n}$ or $b^{2^j t} \equiv -1 \pmod{n}$ for some integer j with $0 \leq j \leq s - 1$. The fact that n is a strong pseudoprime to the base b for at most $(n - 1)/4$ bases with $1 \leq b \leq n - 1$ when n is a composite integer is the basis for a probabilistic primality test which can be used to find extremely large numbers almost certainly prime in just seconds: Pick at random k different positive integers less than n. If n is composite, the probability that n is a strong pseudoprime to all k bases is less than $(1/4)^k$.

The naive way to factor an integer n is to divide n by successive primes not exceeding n. If a prime factor p is found, this process is repeated for n/p, and so on. Factoring large integers n in this way is totally infeasible, requiring prohibitively large times. This shortcoming has led to the development of improved factorization techniques. Techniques including the quadratic sieve, the number field, and the elliptic curve factoring methods have been developed, making it possible to factor numbers with as many as 150 digits, although a large number of computers may be required. However, the factorization of integers with 200 digits is considered far beyond current capabilities, with the best-known methods requiring an astronomically larger time.

Interest in primality testing and factorization has increased since the mid-1970s because of their importance in cryptography. The Rivest-Shamir-Adleman (RSA) public-key encryption system uses the product of two large primes, say with 100 digits each, as its key. The security of the system rests on the difficulty of factoring the product of these primes. The product of the two primes held by each person is made public, but the separate primes are kept secret.

Number Theory

Sieves

In the third century B.C., Eratosthenes showed how all primes up to an integer n can be found when only the primes p up to n are known. It is sufficient to delete from the list of integers, starting with 2, the multiples of all primes up to n. The remaining integers are all the primes not exceeding n. Many problems, including the relative density of the set of twin primes (primes that differ by 2 such as 17 and 19), or the Goldbach conjecture that every even integer greater than 2 is the sum of two primes, have been attacked by sieve methods with partial success. Nevertheless, in spite of improvements of the method, it is not known whether the number of twin primes is finite or infinite or whether Goldbach's conjecture is true or false. (Goldbach's conjecture has been verified by computers up to 4×10^{14}.) In 1966 J. Chen showed that all even integers other than 2 are sums of a prime and another integer that is either prime or a product of only two primes. (The evidence that there are infinitely many twin primes continues to mount. As of 2000, the largest know twin primes were $1{,}693{,}965 \cdot 2^{66{,}443} \pm 1$.)

Primitive roots and discrete logarithms

The order of a modulo m, where a and m are coprime, is the least positive integer x such that formula (5)

$$a^x \equiv 1 \pmod{m} \tag{5}$$

is satisfied. Euler's theorem shows such an integer exists since $a^{\phi(m)} \equiv 1 \pmod{m}$. An integer a is called primitive root modulo m if the order of a modulo m equals $\phi(m)$, the largest possible order. The integer m has a primitive root if and only if $m = 2$, 4, p^k, or $2p^k$ where p is an odd prime.

A solution of the congruence $b^x \equiv c \pmod{n}$ for x is known as a discrete logarithm to the base b of c modulo n. The security of many cryptographic systems is based on the difficulty of finding discrete logarithms. The computational complexity of finding discrete logarithms modulo a prime p is similar to that of factoring an integer n of similar size to p. Several important methods for finding discrete logarithms modulo a prime are based on sieve techniques. The problem of finding discrete logarithms modulo p is much easier when $p - 1$ has only small prime factors.

Quadratic resides and reciprocity

Quadratic residues modulo m are integers that are perfect squares modulo m. More precisely, a is a quadratic residue modulo m if the congruence shown in formula (6)

$$x^2 \equiv a \pmod{m} \tag{6}$$

has a solution for x, where a and m are coprime. If p is an odd prime, the Legendre symbol (a/p) is defined to be +1 if a is a quadratic residue modulo p, and to be -1 if a is a quadratic nonresidue modulo p.

The law of quadratic reciprocity, proved by Gauss, states that if r and q are odd primes, then Eq. (7) is valid. Moreover, Eq. (8)

$$\left(\frac{p}{q}\right)\left(\frac{q}{p}\right) = (-1)^{[(p-1)/2][(q-1)/2]} \tag{7}$$

$$\left(\frac{-1}{p}\right) = (-1)^{(p-1)/2} \tag{8}$$

is true. The Legendre symbol can be generalized to odd composite moduli (Jacobi symbol) and to even integers (Kronecker symbol); reciprocity laws hold for these more general symbols.

Quadratic forms

The expression $F(x, y) = Ax^2 + Bxy + Cy^2$, where A, B, and C are integers and x and y are variables that take on integer values, is called a binary quadratic form with discriminant $B^2 - 4AC$. Two forms $F(x, y) = Ax^2 + Bxy + Cy^2$ and $F_1(x', y') = A_1 x'^2 + B_1 x' y' + C_1 y'^2$ are equivalent if there are integers a, b, c, and d such that $ad - bc = 1$, $x' = ax + by$, and $y' = cx + dy$. Equivalent binary quadratic forms have the same discriminant and represent the same integers. The number of classes of equivalent forms with a given discriminant is finite. There are only finitely many negative discriminants with a given class number, with exactly nine with class number one. In the theory of binary quadratic forms with positive discriminant D, Pell's equation, $x^2 - Dy^2 = 1$, plays a fundamental role.

Number Theory

Sums of squares and similar representations

Pierre de Fermat observed, and Euler proved, that every prime number congruent to 1 modulo 4 is the sum of two squares. J. L. Lagrange showed that every positive integer is the sum of at most four squares, and for some integers four squares are indeed needed. E. Waring conjectured in 1782 that to every positive integer k there is a number $g(k)$ such that every natural number is the sum of at most $g(k)$ kth powers. This was first proved in 1909 by David Hilbert. Further work has been devoted to finding $g(k)$, the least integer such that every positive integer is the sum of at most $g(k)$ kth powers and $G(k)$, the least integer such that every sufficiently large integer is the sum of at most $G(k)$ kth powers. It is known that $G(2) = 4$ and $G(4) = 16$, but the value of $G(3)$ is unknown. Although there is strong numerical evidence that $G(3) = 4$, all that is currently known is that $4 \leq G(3) \leq 7$. New and improved bounds for $G(k)$ for small integers k are established with some regularity such as the bounds $G(19) \leq 134$ established and $G(20) \leq 142$.

Diophantine equations

Single equations or systems of equations in more unknowns than equations, with restrictions on solutions such as that they must all be integral, are called diophantine equations, after Diophantus who studied such equations in ancient times. A wide range of diophantine equations have been studied. For example, the diophantine equation (9)

$$x^2 + y^2 = z^2 \qquad (9)$$

has infinitely many solutions in integers. These solutions are known as pythagorean triples, since they correspond to the lengths of the sides of right triangles where these sides have integral lengths. All solutions of this equation are given by $x = t(u^2 - v^2)$, $y = 2tuv$, and $z(u^2 + v^2)$, where t, u, and v are positive integers.

Perhaps the most notorious diophantine equation is Eq. (10).

$$x^n + y^n = z^n \qquad (10)$$

Fermat's last theorem states that this equation has no solutions in integers when n is an integer greater than 2 where $xyz \neq 0$. Establishing Fermat's last theorem was the quest of many mathematicians over 200 years. In the 1980s, connections were made between the solutions of this equations and points on certain elliptic curves. Using the theory of elliptic curves, A. Wiles completed a proof of Fermat's last theorem based on these connections in 1995.

Algebraic number theory

Attempts to prove Fermat's last theorem led to the development of algebraic number theory, a part of number theory based on techniques from such areas as group theory, ring theory, and field theory. Gauss extended the concepts of number theory to the ring $R[i]$ of complex numbers of the form $a + bi$, where a and b are integers. Ordinary primes $p \equiv 3 \pmod 4$ are also prime in $R[i]$, but $2 = -i(1 + i)^2$ is not prime, nor are primes $p \equiv 1 \pmod 4$ since such primes split as $p = (a + bi)(a - bi)$. More generally, an algebraic number field $R(\theta)$ of degree n is generated by the root of a polynomial equation $f(x) = 0$ of degree n with rational coefficients. A number in this field is called an algebraic integer if it satisfies an algebraic equation with integer coefficients with initial coefficient 1. The algebraic integers in an algebraic number field form an integral domain. But, prime factorization may not be unique; for example, in $R[-5]$, $21 = 3 \cdot 7 = (1 + 2\sqrt{-5}) \cdot (1 - 2\sqrt{-5})$ where each of the four factors in the two products is prime.

Analytic number theory

There are many important results in number theory that can be established by using methods from analysis. For example, analytic methods developed by G. F. B. Riemann in 1859 were used by J. Hadamard and C. J. de la Vallée Poussin in 1896 to prove the famous prime number theorem. This theorem, first conjectured by Gauss about 1793, states that (x), the number of primes not exceeding x, behaves as shown in Eq. (11).

$$x^a + y^b = z^c \tag{11}$$

These methods of Riemann are based on (s), the function defined by Eq. (12),

$$\lim_{x \to \infty} \frac{\pi(x)}{(x/\log x)} = 1 \tag{12}$$

where $s = \sigma + it$ is a complex variable; the series in this equation is convergent for 1. Via an analytic continuation, this function can be defined in the whole complex plane. It is a meromorphic function with only a simple pole of residue 1 at $s = 1$. However, the fundamental theorem of arithmetic can be used to show that this series equals the product over all primes p shown in Eq. (12). It can be shown that $\zeta(s)$ has no zeros for $= 1$; this result and the existence of a pole at $s = 1$ suffice to prove the prime number theorem. Many additional statements about $\pi(x)$ have been proved. Riemann's work contains the still unproved so-called Riemann hypothesis: all zeros of $\zeta(s)$ have a real part not exceeding ½.

Another important result that can be proved by analytic methods is Dirichlet's theorem, which states that there are infinitely many prime numbers in every arithmetic progression $am + b$, were a and b are coprime positive integers.

Diophantine approximation

A real number x is called rational if there are integers p and q such that $x = p/q$; otherwise x is called irrational. The number $b^{1/m}$ is irrational if b is an integer which is not the mth power of an integer (for example, 2 is irrational). A real number x is called algebraic if it is the root of a monic polynomial with integer coefficients; otherwise x is called transcendental. The numbers e and are transcendental. That is transcendental implies that it is impossible to square the circle. See also: Circle; e (mathematics)

The part of number theory called diophantine approximation is devoted to approximating numbers of a particular kind by numbers from a particular set, such as approximating irrational numbers by rational numbers with small denominators. A basic result is that, given an irrational number x, there exist infinitely many fractions h/k that satisfy the inequality (13),

$$\zeta(s) = \sum_{n=1}^{\infty} \frac{1}{n^s} = \prod_p \frac{1}{1-p^{-s}} \tag{13}$$

where c is any positive number not exceeding 5. However, when c is greater than 5, there are irrational numbers x for which there are only finitely many such h/k. The exponent 2 in inequality

(16) cannot be increased, since when x is algebraic and not rational the Thue-Siegel-Roth theorem implies that the inequality (14),

$$\left| x - \frac{h}{k} \right| < \frac{1}{ck^2} \tag{14}$$

where is any number greater than zero (however small), can have at most only finitely many solutions h/k.

In 1851, J. Liouville showed that transcendental numbers exist; he did so by demonstrating that the number x given by Eq. (15)

$$\left| x - \frac{h}{k} \right| < \frac{1}{ck^{2+\epsilon}} \tag{15}$$

has the property that, given any positive real number m, there is a rational number h/k that satisfies Eq. (16).

$$x = \sum_{j=1}^{\infty} 10^{-j} \tag{16}$$

Additive number theory

Problems in additive number theory can be studied using power series. Suppose that $m_1, m_2, \ldots, m_k, \ldots$ is a strictly increasing sequence of positive integers, such as the sequence of perfect squares, the sequence of prime numbers, and so on. If the power series of Eq. (17) is raised to the qth power to yield Eq. (18),

$$\left| x - \frac{h}{k} \right| < \frac{1}{k^n} \tag{17}$$

$$f(x) = \sum_{k=1}^{\infty} x^{m_k} \tag{18}$$

then the coefficients A_j represent the number of times that j can be written as the sum of q elements from the set $\{m_j\}$. If it can be shown that A_k is positive for all positive integers k, then it has been shown that every positive integer k is the sum of q numbers from this set. Function-theoretic methods can be applied, as they have been by C H. Hardy and J. E. Littlewood and by I. M. Vinogradov, to determine the coefficients A_k. For example, if m_k is the kth power of a positive integer and q is sufficiently large, Waring's theorem can be proved. Similarly, if m_k is the kth odd prime, this technique has been used to show that every sufficiently large odd number is the sum of at most three primes.

A partition of the positive integer n is a decomposition of n into a sum of positive integers, with order disregarded. Euler showed that the generating function for the number of partitions $p(n)$ is given by Eq. (19).

$$f(x)^q = \sum_{j=1}^{\infty} A_j x^j \qquad (19)$$

A recurrence formula for $p(n)$ can be derived from this formula, and in 1917 Hardy and S. Ramanujan derived an asymptotic formula for $p(n)$ from it using function-theoretic methods. The first term is formula (20).

$$\sum_{n=0}^{\infty} p(n)x^n = \prod_{m=1}^{\infty}(1-x^m)^{-1} \qquad (20)$$

Ramanujan discovered that the function $p(n)$ satisfies many congruences, such as Eq. (21) and (22).

$$p(n) \approx \frac{\exp(\pi\sqrt{2n/3})}{4n\sqrt{3}} \qquad (21)$$

$$p(5n+4) \equiv 0 \pmod{5} \qquad (22)$$

Euler showed that the number of partitions into odd parts equals the number of partitions into distinct parts. For instance, 7 has five partitions into odd parts (7, 1 + 1 + 5, 1 + 3 + 3, 1 + 1 + 1 + 1 + 3, and 1 + 1 + 1 + 1 + 1 + 1 + 1) and five partitions into distinct parts (7, 1 + 6, 2 + 5, 3 + 4, 1 + 2 + 4). Many additional results have been proved about partitions.

2

Identities for Linear Recurring Sequences

There are several methods for computing the Nth term (mod M) of a linear recurring sequence of order d in $\log_2(N)$ steps, but most such methods require d^2 full multiplications (mod M) per step. The algorithm described below requires only $d(d+1)/2$ multiplications per step.

The algorithm can be described in terms of the basis sequences $B\_j(k)$, defined by

$$x^n = \sum_{j=0}^{d-1} B\_j(n)\, x^j \quad \mod f(x)$$

where $f(x)$ is the characteristic polynomial of the recurrence. There are several useful identities on these sequences. In particular

$B\_k(n1+n2+..+nt+d)$

$\quad = B\_k(a1+a2+..+at+d)\, B\_a1(n1)\, B\_a2(n2)\, ...\, B\_at(nt)$

where summation from 0 to d-1 is implied over each index a*

NOTATION: Summation from 0 to d-1 is implied over any index that appears both as a subscript and as an argument in a single term

Identities for Linear Recurring Sequences   27

For computing the Nth term of a linear recurrence (mod M), the most useful special case of the above identity is

$$B\_k(2n+r) = B\_k(a1+a2+r)\ B\_a1(n)\ B\_a2(n)$$

where r is set to either 0 or 1 according to the binary bits of N. Because of the symmetry of the right hand side, only $d(d+1)/2$ full multiplications are required per step.

There is a nice relation between the basis sequences B and the s(n) sequences (i.e., the sums of the nth powers of the roots of f). Let $\{i,j,..,k\}$ be any given permutation of the integers $\{1,2,..,t\}$ with the disjoint cyclic components c1, c2,..,cr. Then we have

$$B\_a1(n1+ai)\ B\_a2(n2+aj)\ ...\ B\_at(nt+ak) = \prod_{q=1}^{r} s(\ \text{sum}(n[q])\ )$$

where sum(n[q]) denotes the sum over all the n'x' such that x is in cq. In the most trivial case, this identity reduces to the fact that s(n) is the "trace" (or "contraction") of the basis sequences

$$s(n) = B\_a(n+a)$$

Of course, given the basis sequence elements, we can easily compute the Nth terms of any linear recurring sequence (not just s(n) with the characteristic polynomial f(x), since every such sequence is a linear combination of the basis sequences.

3
Half-Totient Tree

The number of ways in which an integer n>2 can be partitioned into two co-prime parts is

$$H(n) = \frac{phi(n)}{2}$$

where phi(n) is Euler's "totient" function, i.e., the number of integers less than and co-prime to n. The half-totient function is intimately involved in many interesting areas of mathematics. For example, H(n) gives the number of distinct linear fractional transformations of order n.

The half-totient function can be used to construct a tree containing all the integers. On the zeroth rank we have just the integers 1 and 2. The immediate "ancestors" of 1 and 2 are

```
1:   3   4   6
2:   5   8  10  12
```

so these numbers constitute the first rank. The ancestors of these numbers constitute the second rank

```
3:    7   9  14  18
4:   15  16  20  24  30
6:   13  21  26  28  36  42
```

Half-Totient Tree

```
 5:  11  22
 8:  17  32  34  40  48  60
10:  25  33  44  50  66
12:  35  39  45  52  56  70  72  78  84  90
```

and so on. Each positive integer is in either the "1 family" or the "2 family", i.e., it ultimately descends to either 1 or 2. Let $T(n)$ denote the "type" of the integer n, where $T(n)=1$ or 2 depending on whether 1 or 2 is a descendant of n.

The values of $T(p^k)$ for the first few primes are tabulated below:

$$k$$

p	1 2 3 4 5 6 7 8 9 ...		p	1 2 3 4 5 6 7 8 9 ...
2	2 1 2 1 2 1 2 1 2...		41	1 2 1 2 1 2 1 2 1....
3	1 1 1 1 1 1 1 1 1...		43	1 1 1 1 1 1 1 1 1....
5	2 2 2 2 2 2 2 2 2...		47	2 1 2 1 2 1 2 1 2...
7	1 1 1 1 1 1 1 1 1...		53	1 2 1 2 1 2 1 2 1...
11	2 1 2 1 2 1 2 1 2...		59	1 1 2 1 2 1 2 1 2...
13	1 2 1 2 1 2 1 2 1...		61	1 2 1 2 1 2 1 2 1...
17	2 2 2 2 2 2 2 2 2...		67	2 1 2 1 2 1 2 1 2...
19	1 1 1 1 1 1 1 1 1...		71	2 1 2 1 2 1 2 1 2...
23	2 1 1 1 1 1 1 1 1...		73	1 2 1 2 1 2 1 2 1...
29	1 2 2 2 2 2 2 2 2...		79	2 1 2 1 2 1 2 1 2...
31	1 1 1 1 1 1 1 1 1...		83	1 1 1 1 1 1 1 1 1...
37	1 1 1 1 1 1 1 1 1...		89	2 2 2 2 2 2 2 2 2...
			97	2 2 2 2 2 2 2 2 2...

This clearly suggests that the powers of any given prime are either all type 1, all type 2, or alternating between 1 and 2. For the primes less than 100, only the first powers of 23, 29, and 59 violate their respective patterns.

I'm interested in whether, for any given integer N, the type of N can be expressed simply in terms of the types of the prime factors of N. In other words, I'm trying to find "equivalence classes" relative to "type" under multiplication. The best classification I've been able to find is summarized below:

U_p < 0				U_p = 0		U_p > 0			
U_p odd		U_p even		U_p even		U_p odd		U_p even	
-1	+1	-1	+1	-1	+1	-1	+1	-1	+1
3	109	379	1949	7	37	11	13	23	5
19	653	487	2053	43	229	47	41	31	17
127	757	1999	2269	223	1297	59	53	83	29
163	3889	2287	2917	271	1549	67	61	103	89
811	4549	2647	11689	1307	1597	71	73	107	97
883	4789	3079	12637	1423	1621	79	137	131	101
1459	4861	11827	13693	1483	7793	167	193	139	113
3919	5869	12043	13933	1567	7993	179	233	151	149
3943	23473	12779	15877	1627	8209	227	241	191	157

etc.

where

$$U\_n = s2(n) - 2\, s3(n)$$

and sx(n) is the sum of the highest exponents k_j of x dividing phi(n_j) for the sequence of n values given by $n\_0 = n$

$$n\_(j+1) = phi(n\_j)/(x^{\wedge}(k\_j))$$

The values of "-1" and "+1" in the table headers signify the residue class of p (mod 4).

Notice that all the primes in a given column exhibit the same pattern of types for consecutive powers. For example, the primes 11, 47, 59, 67, 71, 79, etc all exhibit the pattern "212121..." for the types of their consecutive powers.

The rough "equivalence" of primes in each of the ten columns extends to the types of products of different primes as well. For example, the product of any two primes from the column 5,17,29,89... is usually of type 2. The only exception to this for primes less than 1000 is the product of 13 and 509, which is type 1. (The prime 509 happens to be the smallest prime whose "power pattern" is exceptional in the SECOND power, viz, we expect "12121212..." but it is actually "1112121212...".)

The product of any two DISTINCT primes from the column

Half-Totient Tree

23,31,83... is of type 2, without exception for primes less than 1000, although the square of any prime in this column is type 1.

As can be seen in the above table, the great majority of primes have $U\_p > 0$, so they fall in one of the four right-most columns. If we label these four columns as A, B, C, and D, then we can express a rough generalization for the products of any of these primes as follows:

If, for a given integer N, the symbols a,b,c,d denote the sums of the prime exponents of primes of type A, B, C, and D respectively, then N is type 2 iff a+b is even, unless (b=c=d=0 and a is even) or (a=b=c=0), in which case N is type 1.

Caution: There are exceptions to this rule, such as the first or second powers of certain primes noted earlier, as well as some composite exceptions.

It's also interesting to consider the possible ranges of integers of a certain rank. It's not too hard to show that the largest integer of rank r must be at least $12(7^{(r-1)})$. Obviously the smallest member of rank r can be no smaller than about 2^r. The smallest integer of rank r is often related to Cunningham chains, which give the slowest rate of "descent" under the half-totient function, i.e., decreasing by a factor of 2 on each step.

Another interesting aspect is to consider the ratio of the number of type 1 integers to number of type 2 integers less than x as x -> inf.

Here is a brief table showing the number of types 1 and 2 on each rank:

Rank	# of type 1	# of type 2	total
0	1	1	2
1	3	4	7
2	15	23	38
3	84	142	226
4	439	860	1299
5	2371	4850	7221
6	12779	26614	39393
7	68638	145947	214585
8	366344	791667	1158011
9	1954833	4275444	6230277

This shows that on any given rank the number of type 2 integers exceeds the number of type 1 integers. However, the type 1 integers always outnumber the type 2 integers less than any fixed N. This is because the type 1 integers are more prevalent in the lower members of each rank, and before any rank is completely filled the next rank has started to appear. Since each rank has about 5 times as many members as the previous rank, the ratio of types up to any fixed N is always dominated by the low members of the nascent rank. If we let Type1(x) and Type2(x) denote, respectively, the numbers of type 1 and 2 integers less than x, we can plot the proportion of the two types as shown in the figure below.

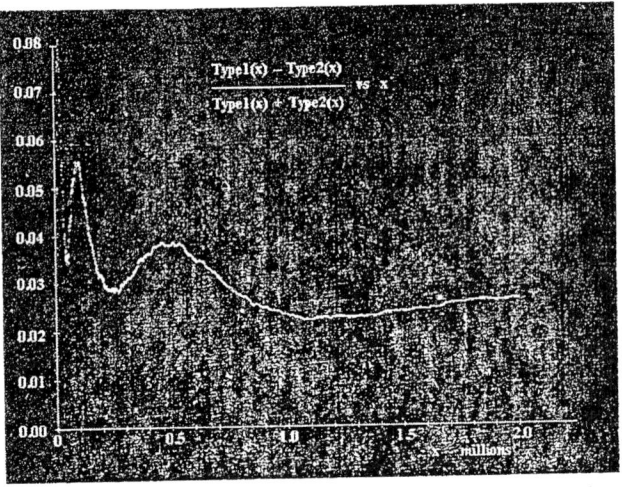

The logarithmic periodicity of this proportion is shown even more clearly in the following plot as a function of the log of x

Half-Totient Tree

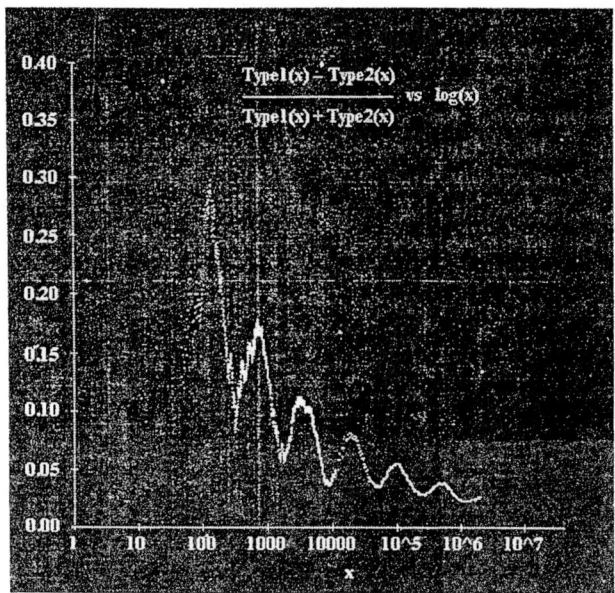

Does this ratio converge on 0? If not, what is the asymptotic value?

For another example of this kind of "logarithmic wave" propogating through the natural numbers.

4

Limit Cycles of xy (mod x+y)

For any two positive integers x and y define $F(x,y) = xy$ (mod x+y). Given two initial values $x[0]$ and $x[1]$ we can form a sequence using the recurrence

$$x[n] = F(\ x[n-1]\ ,\ x[n-2]\ )$$

For some initial values this produces a trivial outcome. For example, if $x[1] = x[0]$ is odd, then $x[n] = x[0]$ for all n. Slightly less obvious is the fact that if $x[0]=2k$ and $x[1]=2k-6$ then $x[n]=2k-6n$ for all n up to k/3, at which point the sequence terminates with a value of 0.

Most initial values lead to a fixed point, but some lead to a limit cycle. The smallest limit cycle is {5,7,11}. Of course, it follows that {7,5,11} is also a limit cycle. (I wonder if it's true that if $xy = z$ (mod x+y) and $xz = y$ (mod x+z) then $yz = x$ (mod y+z).)

Anyway, here are several examples of limit cycles with period 3 (along with their gcd factorizations):

{5,7,11}	1*[5,7,11]
{69,99,111}	3*[23,33,37]
{87,111,153}	3*[29,37,51]
{184,704,776}	8*[23,88,97]
{125,475,575}	25*[5,19,23]
{384,864,1056}	96*[4,9,11]

Limit Cycles of xy (mod x+y)

{315,525,735}   105*[3,5,7]
{324,756,864}   108*[3,7,8]
{1575,2205,2835}   45*[35,49,63]

The smallest limit cycles with period 4 are

{96,304,384,464}   16*[6,19,24,29]
{128,192,256,320}   64*[2,3,4,5]
{243,1701,1215,2187}   243*[1,7,5,9]
{495,1375,1815,1045}   55*[9,25,33,19]
{1331,1815,2783,2541}   121*[11,15,23,21]

There smallest examples with period 5 are

{124,310,248,434,558}   62*[2,5,4,7,9]
{243,621,567,549,675}   9*[27,69,63,61,75]
{392,490,686,980,882}   14*[4,5,7,10,9]

Here are limit cycles with periods 6 and 7:

{23,61,59,119,79,95}   1*[23,61,59,119,79,95]

{77,343,371,161,147,259,315}   7*[11,49,53,23,21,37,45]

The table below presents one cycle for for each of the periods 8 through 19, and also a cycle of period 22.

T=8, gcd=59		T=9, gcd=65		T=10, gcd=441		T=11, gcd=319	
1711	29	455	7	3087	7	15631	49
6431	109	1235	19	10143	23	10527	33
3599	61	845	13	9261	21	13717	43
5959	101	1495	23	18963	43	1595	5
7847	133	2015	31	6615	15	13079	41
13157	223	845	13	5733	13	9251	29
8319	141	975	15	3087	7	9889	31
11387	193	1235	19	4851	11	13079	41
		1885	29	3969	9	5423	17
				8379	19	9231	29
						12441	39

36 Number Theory

T=12, gcd=1023		T=13, gcd=71		T=14, gcd=7161		T=15, gcd=107	
17391	17	35855	505	279279	39	17441	163
25575	25	29749	419	136059	19	13375	125
33759	33	60563	853	393855	55	27071	253
17391	17	54599	769	494109	69	2033	19
3069	3	32731	461	221991	31	28783	269
13299	13	46079	649	565719	79	27071	253
9207	9	24779	349	708939	99	21293	199
11253	11	56587	797	594363	83	20651	193
17391	17	70361	991	451143	63	22791	213
5115	5	47783	673	737583	103	6313	59
11253	11	35855	505	93093	13	18511	173
9207	9	18673	263	136059	19	13375	125
		25631	361	221991	31	21721	203
				207669	29	28783	269
						6527	61

T=16, gcd=12051		T=17, gcd=75087		T=18, gcd=551		T=19, gcd=4875	
445887	37	2027349	27	143811	261	375375	77
397683	33	1276479	17	330049	599	687375	141
735111	61	3078567	41	15979	29	443625	91
1000233	83	2628045	35	40223	73	1038375	213
1409967	117	675783	9	53447	97	531375	109
735111	61	2177523	29	72181	131	960375	197
2012517	167	375435	5	73283	133	541125	111
1554579	129	2477871	33	132791	241	258375	53
2374047	197	1877175	25	96425	175	609375	125
1988415	165	3679263	49	137199	249	102375	21
1072539	89	1426653	19	220951	401	589875	121
277173	23	1877175	25	82099	149	24375	5
735111	61	976131	13	192299	349	453375	93
397683	33	1276479	17	66671	121	180375	37
1000233	83	2027349	27	197809	359	316875	65
60255	5	3078567	41	93119	169	424125	87
		976131	13	251807	457	180375	37
				291479	529	258375	53
						365625	75

T=22, gcd=1129

524337	553
649175	575
1136903	1007
495631	439

Limit Cycles of xy (mod x+y)

567887	503
205961	209
134351	119
296927	263
86933	77
89191	79
134351	119
154673	137
224671	199
351119	311
147899	131
339829	301
486599	431
473051	419
655949	581
712399	631
1096259	971
716915	635

It can be shown that there exists a cycle of length n for every positive integer n. See More Results on the Form xy (mod x+y).

There is an interesting class of cycles of period 9, in which the cycle consists of just four distinct numbers in the pattern A BC A DB A CD. Several examples are shown below

```
3575  9295  12155  3575  7865  9295  3575  12155  7865       5 11 13
  5    13     17     5    11    13     5     17    11

8381  5423  7395  8381  9367  5423  8381  7395  9367         17 29
 17    11    15    17    19    11    17    15    19

8300  5810  9130  8300  10790  5810  8300  9130  10790       2 5 83
 10     7    11    10     13     7    10    11    13

52731 29295 41013 52731 76167 29295 52731 41013 76167        3 3 3 7 31
  9     5     7     9    13     5     9     7    13

72471 123627 46893 72471 89523 123627 72471 46893 89523      3 7 7 29
 17    29     11    17    21    29     17    11    21

132759 122925 113091 132759 34419 122925 132759 113091 34419  3 11 149
  27    25     23     27     7    25     27     23     7

106029 82467 129591 106029 223839 82467 106029 129591 223839  3 3 7 11 17
  9      7     11      9     19      7     9      11    19
```

| 197519 | 245891 | 116899 | 197519 | 173333 | 245891 | 197519 | 116899 | 173333 | 29 139 |
| 49 | 61 | 29 | 49 | 43 | 61 | 49 | 29 | 43 | |

931095	517275	1344915	931095	1138005	517275	931095	1344915	1138005	3
3 5 11 11 19									
9	5	13	9	11	5	9	13	11	

1576575	735735	1156155	1576575	1366365	735735	1576575	1156155	1366365	
3 5 7 7 11 13									
15	7	.1	15	13	7	15	11	13	

In many of these cycles the reduced term appearing three times (i.e., the "A" term of the cycle) is the product of the first two or three prime divisors of the cycle's greatest common divisor.

Another interesting cycle of period 9 is shown below. In this cycle the greatest common divisor of the cycle is one of the terms of the cycle.

| 80087 | 51821 | 98931 | 80087 | 108353 | 4711 | 80087 | 23555 | 61243 | 7 673 |
| 17 | 11 | 21 | 17 | 23 | 1 | 17 | 5 | 13 | |

A Knot of Congruences

Problem:

Prove (or disprove?) that the only solutions of

$$ab = c \pmod{a+b}$$
$$ac = b \pmod{a+c}$$
$$bc = a \pmod{b+c}$$

in positive coprime integers a,b,c are {1,1,1} and {5,7,11}.

Discussion:

Note that if a,b,c are not required to be positive then there are many coprime solutions, such as {299,-49,-401}. Also, if a,b,c are positive but not necessarily coprime, then there are many solutions, such as
{69,99,111}.

Since the congruences are symmetrical we can assign the parameter labels so that a < b < c. In explicit form the system of congruences can be written as

Limit Cycles of xy (mod x+y)

$$ab = c + x(a+b)$$
$$ac = b + y(a+c)$$
$$bc = a + z(b+c)$$

where x,y,z are integers. These equations imply that

$$\left(\frac{a-1}{z}\right)\left(\frac{b-1}{y}\right)\left(\frac{c-1}{x}\right) - \left(\frac{a-1}{z}\right) - \left(\frac{b-1}{y}\right) - \left(\frac{c-1}{x}\right) = 2$$

An equation of this form, i.e., ABC-A-B-C = 2, has the positive integer solutions {2,2,2}, {1,2,5}, and {1,3,3}, and the integer solution {-1,-1,-1}. However, not all triples {a,b,c} that satisfy the original set of congruences lead to integer quantities in the brackets.

Another interesting implication of the three coupled equations is

$$\left(\frac{c-a}{by}\right) = \frac{\left(\frac{c-b}{az}\right) + \left(\frac{b-a}{cx}\right)}{1 + \left(\frac{c-b}{az}\right)\left(\frac{b-a}{cx}\right)}$$

which shows that the "normalized" distances [a to b] and [b to c] add up to the total distance [a to c] in accord with the addition rule for velocities in special relativity. (This can be seen most clearly by setting x=1/c, y=1/b, and z=1/a.)

For any integer solution A,B,C of the original congruences, define

$$g = GCD(A,B,C) \quad a = A/g \quad b = B/g \quad c = C/g$$

and put M = LCM(a+b,a+c,b+c). It's clear that a,b,c are pairwise coprime and that infinitely many other solution triples are given by

$$A' = a(Mk+g) \quad B' = b(Mk+g) \quad C' = c(Mk+g)$$

where k is any integer. A solution {ag,bg,cg} with g < M is called a minimal solution.

Examination of the minimal solutions with g=1 shows that certain values of (a+b+c) occur frequently. For example, the following triples all have (a+b+c) = -361

a	b	c
47	-53	-355
109	-169	-301
119	-131	-349
143	-145	-359
149	-121	-389
1619	-179	-1801

The basic equations imply that if a,b,c are coprime then a divides xy-1, b divides xz-1, and c divides yz-1, but this doesn't seem to be sufficient to prove that {1,1,1} and {5,7,11} are the only positive coprime solutions. It appears that for most (all?) g > 1 there are infinitely many minimal solutions, but I can't prove that either.

5

Rounding Up To PI

Beginning with any positive integer n, round up to the nearest multiple of n-1, then up to the nearest multiple of n-2, and so on up to the nearest multiple of 1. Let f(n) denote the result. For example, f(10)=34. Interestingly, the ratio $n^2/f(n)$ approaches pi (i.e., 3.14159...) as n increases.

To derive this limit it's useful to examine the sequence of numbers that are produced as we "round up" from, say, 100. In the following table x is the "rounding modulus" and y is the resulting rounded value for each step of the algorithm in the computation of f(100). The value of w is just y/x.

x	w	y	x	w	y	x	w	y	x	w	y
100	1	100	75	26	1950	50	51	2550	25	122	3050
99	2	198	74	27	1998	49	53	2597	24	128	3072
98	3	294	73	28	2044	48	55	2640	23	134	3082
97	4	388	72	29	2088	47	57	2679	22	141	3102
96	5	480	71	30	2130	46	59	2714	21	148	3108
95	6	570	70	31	2170	45	61	2745	20	156	3120
94	7	658	69	32	2208	44	63	2772	19	165	3135
93	8	744	68	33	2244	43	65	2795	18	175	3150
92	9	828	67	34	2278	42	67	2814	17	186	3162
91	10	910	66	35	2310	41	69	2829	16	198	3168
90	11	990	65	36	2340	40	71	2840	15	212	3180
89	12	1068	64	37	2368	39	73	2847	14	228	3192

x	w	y	x	w	y	x	w	y	x	w	y
88	13	1144	63	38	2394	38	75	2850	13	246	3198
87	14	1218	62	39	2418	37	78	2886	12	267	3204
86	15	1290	61	40	2440	36	81	2916	11	292	3212
85	16	1360	60	41	2460	35	84	2940	10	322	3220
84	17	1428	59	42	2478	34	87	2958	9	358	3222
83	18	1494	58	43	2494	33	90	2970	8	403	3224
82	19	1558	57	44	2508	32	93	2976	7	461	3227
81	20	1620	56	45	2520	31	96	2976	6	538	3228
80	21	1680	55	46	2530	30	100	3000	5	646	3230
79	22	1738	54	47	2538	29	104	3016	4	808	3232
78	23	1794	53	48	2544	28	108	3024	3	1078	3234
77	24	1848	52	49	2548	27	112	3024	2	1617	3234
76	25	1900	51	50	2550	26	117	3042	1	3234	3234

The rounding algorithm essentially states that, on each step, y must be the least integer that equals or exceeds the previous y and is divisible by x. Notice that initially the ratio w increases by 1 on each step, so y can be expressed as

$$y = (101 - x) x \qquad (1)$$

This is a parabola, and the value of y increases as x decreases until y reaches a maximum of y=2550 at x=50. As x continues to decrease we must now increase the ratio w by 2 on each step in order for the values of y to keep increasing. In other words, after riding the parabola (1) to its maximum we have to shift to the intersecting parabola

$$y = (151 - 2x) x$$

which we can ride up to its maximum of $[151/(2*2)] = 38$. At that point the process shifts to the intersecting parabola

$$y = (189 - 3x) x$$

and so on. In general, the equation for the kth parabola is of the form

$$y = (A\_k - kx) x$$

Rounding Up To PI

where A_k is a constant. Setting the derivative of this expression to zero gives the coordinates $[x_k, y_k]$ of the maximum point on the kth parabola:

$$x_k = \frac{A_k}{2k} \qquad y_k = k(x_k)^2$$

Also, the value of A_k can be inferred from from the coordinates of the intersection of the kth parabola with the maximum point of the previous parabola, which gives

$$A_k = \frac{y_{(k-1)} + k(x_{(k-1)})^2}{x_{(k-1)}}$$

Substituting $(k-1)(x_{(k-1)})^2$ for $y_{(k-1)}$ into this expression and then putting this A_k into the expression for x_k gives

$$x_k = \left(\frac{2k-1}{2k}\right) x_{(k-1)}$$

with the exception that we have $x_1 = (x_0 + 1)/2$ instead of $(x_0)/2$, because we have the initial condition $y_0 = x_0$ rather than $y_0 = k(x_0)^2$. Thus, beginning with $x_0 = y_0 = n$, the "rounding up" algorithm in continuous form yields the following convergent product for x_k

$$x_k = (n+1) \prod_{j=1}^{k} \frac{2j-1}{2j}$$

from which it follows that

$$y_k = k(x_k)^2 = k(n+1)^2 \left(\prod_{j=1}^{k} \frac{2j-1}{2j}\right)^2$$

Since y_k approximates $f(n)$ in the limit as k goes to infinity, this relation implies

$$\frac{(n+1)^2}{f(n)} = \lim_{k \to \inf} \frac{1}{k} \left(\prod_{j=1}^{k} \frac{2j}{2j-1} \right)^2$$

which is equivalent to Wallis's infinite product for PI.

Here's a plot showing the first four parabolas that appear in the algorithm for f(100).

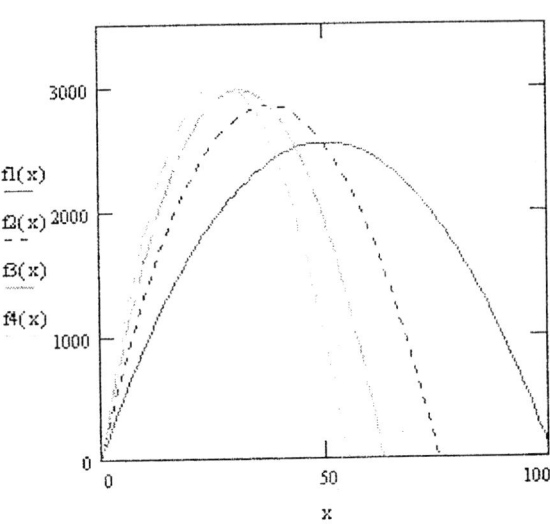

Incidentally, the sequence f(1), f(2), f(3)... can also be derived using a seive method similar to the seive of Eratosthenes. Starting with a list of the natural numbers, delete every 2nd element beginning with the 3rd. This leaves the sequence

1 2 4 6 8 10 12 14 16 18 20 22 24 26 28 30 32 34 ...

From this, delete every 3rd element beginning with the 5th, which leaves

Rounding Up To PI

1 2 4 6 10 12 16 18 22 24 28 30 34 ...

From this, delete every 4th element beginning with the 7th, which leaves

1 2 4 6 10 12 18 22 24 30 34 ...

From this, delete every 5th element beginning with the 9th, which leaves

1 2 4 6 10 12 18 22 30 34 ...

In general, from the kth sequence delete every (k+1)th element beginning with the (2k+1)th element. Notice that this yields the same sequence $f(1), f(2), f(3),...$ as was produced by the "rounding up" algorithm. (Is this equivalence obvious?)

This sequence appears as sequence #377 in Sloane's Handbook of Integer Functions, which refers to a paper by Yosef David in vol 11 of Riveon Lematematika (1957). This paper says that Jabotinski and Erdos proved that

$$f(n) = n^2/pi + O(n^{(4/3)})$$

consistent with our observation on the rounding sequence. There are also some relevant references in Richard Guy's "Unsolved Problems in Number Theory" under the heading of "lucky numbers". In all of these references the authors derive this sequence using the seive method. No one seems to have noticed the simple "rounding up" algorithm.

By the way, notice how easy it would be to mistake the sequence f(n) for the cumulative sums of the Euler totient function phi(n):

1 2 4 6 10 12 18 22 28 32 ...

Is this similarity significant?

6

Fermat's Last Theorem for Cubes

Before considering the integer equation $x^3 + y^3 = z^3$, it's worthwhile to briefly review the simple Pythagorean equation $x^2 + y^2 = z^2$. For primitive solutions we can assume x,y,z are pairwise coprime, x is odd and y is even. The usual approach is to re-write the equation as

$$\left(\frac{y}{2}\right)^2 = \left(\frac{z+x}{2}\right)\left(\frac{z-x}{2}\right)$$

Then, since the two integer factors on the right are coprime (and since we have unique factorization for integers), they must each individually be squares, so we have $z+x = 2u^2$ and $z-x = 2v^2$ for coprime integers u,v, (one odd and one even) from which it follows that $z = u^2 + v^2$, $x = u^2 - v^2$, and $y = 2uv$.

However, there is another approach to solving the Pythagorean equation that makes use of some deeper properties of integers known to Fermat, and that can be generalized to the case of cubes. This alternative approach relies on the fact that numbers of the form $X^2 + Y^2$ with $\gcd(X,Y)=1$ can be "factored" uniquely into a product of primes of the same form, and that the representations of composites of this form are generated by applying the identity

Fermat's Last Theorem for Cubes   47

$$(a^2 + b^2)(c^2 + d^2) = (ac +- bd)^2 + (ad -+ bc)^2$$

It's been speculated that Diophantus knew this identity, although he didn't give it explicitly in any of the (surviving) books of "Arithmetica". The first known explicit description was by Abu Jafar al-Khazin (circa 950 AD), and it also appears in Fibonacci's "Liber Quadratorum" (1225 AD). One could argue that this was really the first discovery of complex numbers, in the abstract sense of Hamilton's ordered pairs, because in C the product of (a,b) and (c,d) is (ac-bd,ad+bc). In any case, Fermat knew that only primes of the form 4k+1 are expressible as a sum of two coprime squares, and those are expressible in only one way. This, combined with the fact that representations of composites are given by the above formula applied to the representations of their factors, enables us to say that if $x^2 + y^2$ is a square then the components x,y are given by squaring a number of the form $(u^2 + v^2)$ using the above identity. As a result we have

$$x^2 + y^2 = (u^2 + v^2)^2 = (u^2 - v^2)^2 + (2uv)^2$$

which of course agrees with our previous solution. Thus, given the theorems about sums of two squares and their unique factorizations that were known to Fermat, this is (arguably) an even more direct solution than the original one, which is perhaps not surprising, since it is essentially employing the field of Gaussian integers, in disguised form.

Now let's consider the analagous equation for cubes, i.e., we seek all non-trivial integer solutions of $x^3 + y^3 = z^3$. Again we consider only primitive solutions, so without loss of generality we can assume x,y,z are coprime, one even and two odd. Changing signs if necessary we can make x and y odd and z even. Now we define x=u+v and y=u-v where (u,v)=1, and u,v have opposite parity.

Substituting into $x^3 + y^3 = z^3$ gives

$$(2u)(u^2 + 3v^2) = z^3 \qquad (1)$$

Since z is even, u must be even and v must be odd. Now

we'll consider two cases. First, assume z is not divisible by 3. In this case 2u is coprime to $u^2 + 3v^2$, so both of those factors must be cubes. Thus we have integers coprime m,n such that

$$u = 4m^3 \qquad (2)$$

$$u^2 + 3v^2 = n^3 \qquad (3)$$

In the case of the Pythagorean equation we had a sum of two squares equal to a square, whereas in this case we have a slightly different quadratic form, $X^2 + 3Y^2$, equal to a cube. Notice that we can't simply subtract u^2 from both sides of (3) and then factor the right hand side, because it is inhomogeneous, i.e., we would have a cube minus a square, which doesn't factor algenraically over the integers. We can, however, proceed to use the second approach, based on factoring the left hand side of (3) into divisors of the same form, provided we know enough about numbers of the form $X^2 + 3Y^2$.

Happily, it turns out that we have a direct analog for the "Fibonacci identity". In fact, for ANY integer N we have

$$(a^2 + Nb^2)(c^2 + Nd^2) = (ac +\!\!- Nbd)^2 + N(ad -\!\!+ bc)^2$$

so we can always multiply together two numbers of the quadratic form $X^2 + NY^2$ to give another number of the same form. With k=3 this identity is

$$(a^2 + 3b^2)(c^2 + 3d^2) = (ac +\!\!- 3bd)^2 + 3(ad -\!\!+ bc)^2$$

With this identity in mind, we state and prove several facts about numbers of the quadratic form $X^2 + 3Y^2$ which are useful for continuing our search for solutions of $x^3 + y^3 = z^3$.

LEMMA 1: Every prime p of the form 3k+1 divides some integer of the form $a^2 + 3b^2$ with (a,b)=1.

PROOF: Since $u^2 + uv + v^2$ is an equivalent form under

Fermat's Last Theorem for Cubes 49

the substitution $u=b+a$ and $v=b-a$, we need only prove that p divides such an integer, with $(u,v)=1$. Consider

$$u^{3k} - v^{3k} = (u^k - v^k)(u^{2k} + u^k v^k + v^{2k})$$

where $3k = p-1$. Setting $v=1$ ensures $(u,v)=1$ and enables us to write

$$u^{3k} - 1 = (u^k - 1)(u^{2k} + u^k + 1)$$

The left hand side is divisible by p according to Fermat's Little Theorem for any integer u coprime to p. Therefore, the right side is also divisible by p for every such u. In order for p to NOT divide any of the number $u^{(2k)} + u^k + 1$, it must divide EACH of the numbers $u^k - 1$ for $u = 1,2,3,...,p-1$. However, the congruence $u^{[(p-1)/3]} = 1 \pmod{p}$ can have no more than $(p-1)/3$ distinct roots, so it is NOT satisfied for 2/3 of the residues modulo p.

Therefore, each of those non-roots is a value of u for which p must divide $u^{(2k)} + u^k + 1$. Also, since more than half of those residues qualify, we can choose an odd u, and then $a = (u-1)/2$ and $b = (u+1)/2$. With these values, p divides $a^2 + 3b^2$, which completes the proof of Lemma 1.

LEMMA 2: If N is an integer of the form $a^2 + 3b^2$, and if the prime $p = c^2 + 3d^2$ divides N, then there exist integers u,v such that $N/p = u^2 + 3v^2$ and the repesentation of N is given by evaluating the product $(p)(N/p) = (u^2 + 3v^2)(c^2 + 3d^2)$ using Fibonacci's formula.

PROOF: Since p divides N, it must divide $Nd^2 - pb^2$. Also, we have

$$Nd^2 - pb^2 = (a^2 + 3b^2)d^2 - (c^2 + 3d^2)b^2$$
$$= (ad + bc)(ad - bc)$$

which shows that the prime p must divide either ad+bc or ad-bc.

Now, we can also write

$$Np = (ac +- ?bd)^2 + 3(ad -+ bc)^2$$

Depending on whether p divides ad+bc or ad-bc, we can choose the sign in the above expression so that p divides the right-most term. Then, since it also divides Np, it must divide the first term on the right. Therefore, dividing the above expression for Np by p^2, we have $N/p = u^2 + 3v^2$ where u,v are the integers given by

$$u = (ac +- 3bd)/p \qquad v = (ad -+ bc)/p$$

again with the choice of sign such that p divides ad-+bc. Solving these two equations for a and b gives

$$a = (cu + 3dv) \qquad b = +-(du - cv)$$

This shows that the representation of N is given by applying Fibonacci's formula to multiply (p)(N/p), which completes the proof of Lemma 2.

LEMMA 3: If we let [n\p] equal +1 or -1 accordingly as n is or is not a square (mod p), and if m,n are residues coprime to p, then [mn\p] = [m\p][n\p].

PROOF: If m,n are both squares (mod p), then obviously mn is also a square. Also, if one of m,n is a square and the other is not, then it follows that their product mn is not, because if $m=x^2$ and $mn=y^2$ we would have $n = (y/x)^2$, contrary to assumption that n is not square. The leaves only the case when neither m nor n is a square. To resolve this case, note that the non-zero multiplication table (modulo p) has unique inverse, so each non-zero residue appears in row and column precisely once. Also, since $x^2=y^2$ (mod p) implies (x-y)(x+y) mod p, it's clear that the squares of the residues 1 through (p-1)/2 are all distinct, and respectively equal to the squares of the residues (p+1)/2 to p-1. Therefore, the squares and non-squares each make up exactly half the non-zero residues. Also, each residue appears p-1 times in the table, so if fill in all the products of two squares, and all the products of a square and a non-square, we are left only with squares, which must be placed in the remaining openings, the products of two non-squares. Therefore [mn\p] = [m\p][n\p], completing the proof of Lemma 3.

Fermat's Last Theorem for Cubes

LEMMA 4: If the integer N is representable in the form $a^2 + 3b^2$ with $(a,3b)=1$, then the only odd prime factors of N are of the form $p = 3k+1$.

PROOF: If N was divisible by a prime p, then we have $a^2 = -3b^2 \pmod{p}$, which implies that (-3) is a square modulo p. It's easy to show that $[-1\backslash p] = (-1)^{\wedge}(p-1)/2$, and by quadratic reciprocity we also have $[3\backslash p] = [p\backslash 3](-1)^{\wedge}(p-1)/2$. From Lemma 3 and quadratic reciprocity it follows that $[-3\backslash p] = [-1\backslash p][3\backslash p] = [p\backslash 3]$. Thus any number of the form $a^2 + 3b^2$ with $(a,3b)=1$ is divisible by only primes of the form $3k+1$, which completes the proof of Lemma 4.

Notes

1. It's possible to avoid the use of full quadratic reciprocity here, but I wonder if Fermat might have just assumed it?

2. If $a^2 + 3b^2$ with $(a,b)=1$ is even, then a,b are odd, in which case either a+b or a-b must be divisible by 4. With that choice of sign we can set $B=a+-b$ and $A=a-+3b$ and then we have $A^2 + 3B^2 = [a^2+3b^2]/4$. Repeating if necessary, we can factor out all powers of 2, leaving an odd proper representation.

LEMMA 5: Every prime p of the form $3k+1$ is expressible in the form $u^2 + 3v^2$ with $(a,b)=1$ in precisely one way.

PROOF: By Lemma 1 we know that p divides some integer of the form $a^2 + 3b^2$. Also, by replacing a and b with their least magnitude residues modulo p, the result is still divisible by p, but now we are assured that a and b are each less than or equal to $(p-1)/2$, from which it follows that $a^2 + 3b^2$ is strictly less than p^2. Therefore, all the prime divisors of $a^2 + 3b^2$ other than p are strictly smaller than p, and according to Lemma 4 all of those prime divisors are of the form $3k+1$, and according to Lemma 3 they are all of the form $u^2 + 3v^2$. Therefore, we can apply Lemma 2 to each of these smaller prime divisors in turn, yielding a unique quotient of the form $a^2 + 3b^2$, until arriving at p. This completes the proof of Lemma 5.

LEMMA 6: The general primitive solution in integers of the equation $x^2 + 3y^2 = N^3$ for odd N is given by $x = u(u^2 - 9v^2)$ and $y = 3v(u^2 - v^2)$ where u,v are coprime integers.

PROOF: By Lemma 4 we know that N^3 is a product of primes of the form $3k+1$, each of which by Lemma 5 has a unique proper representation of the form $a^2 + 3b^2$. Hence by Lemma 2 we can factor $x^2 + 3y^2$ uniquely into a product of primes of this form, and the representation of N^3 is given by applying the Fibonacci product formula. Also, it's easy to verify that Fibonacci multiplication is commutative, in the sense that the two representations given by AB are the same as the two given by BA. Also, we can verify that Fibonacci multiplication is associative, i.e., $(AB)C = A(BC)$, by noting the results

$$[(a^2 + 3b^2)(c^2 + 3d^2)](e^2 + 3f^2)$$
$$= [ace + s1\, 3bde + s2\, 3adf - s1\, s2\, 3bcf]^2$$
$$+ 3[ade - s1\, bce - s2\, acf - s1\, s2\, 3bdf]^2$$

Since both components are squared, we need consider only the magnitudes of the components, so we can multiply each term of the second component by $-s1\, s2$ and write the two components as shown below

$$ace + 3[\ s1\, bde + s2\, adf - s1\, s2\, bcf\ ]$$
$$3\, bdf\ +\ s1\, acf + s2\, bce - s1\, s2\, ade\ ]$$

Notice that the rows transpose (a,b), (c,d), and (e,f), so they have the same symmetry, and if we define $s3 = -s1\, s2$ we have the three-way symmetry

$$s1\, s2 = -s3 \qquad s1\, s3 = -s2 \qquad s2\, s3 = -s1$$

Consequently, the set of proper representations given by the Fibonacci product of three proper representations is the same, regardless of the order in which the product is evaluated.

Furthermore, the number of distinct proper representations of a number equals $2^{(k-1)}$ where k is the number of distinct prime divisors, because we have two proper choices of sign when multiplying two distinct factors (whereas we have no proper choices when multiplying powers of a single prime). Since the number of distinct prime divisors of N is the same as the number of distinct prime divisors of N^3, we can produce all 2^k representations of N^3 as the cubes of the 2^k representations

Fermat's Last Theorem for Cubes 53

of N. Thus, for some coprime integers u,v we have not only

$$(u^2 + 3v^2)^3 = x^2 + 3y^2$$

but also expanding the left side by the Fibonacci formula (which gives a unique *proper* result when cubing a single representation) we have

$$x = u(u^2 - 9v^2) \qquad y = 3v(u^2 - v^2)$$

completing the proof of Lemma 6.

Now (finally!) we can return to our original problem. Recall that on the assumption of the existence of integers x,y,z such that $x^3 + y^3 = z^3$, and assuming first that z is not divisible by 3, we had shown the existence of integers m,n and coprime integers u,v such that

$$u = 4m^3 \qquad (2)$$

$$u^2 + 3v^2 = n^3 \qquad (3)$$

where n is odd. It follows from Lemma 6 that u and v can be expressed in terms of integers r,s as follows

$$u = r^3 - 9rs^2 \qquad v = 3r^2s - 3s^3$$

$$= r(r-3s)(r+3s)$$

Also, since v is odd and u is even, we must have r even and s odd. Further, since $u = 4m^3$, it's clear that r is 4 times a cube, and both r-3s and r+3s are cubes. Thus we have

$$r = 4A^3 \qquad r-3s = B^3 \qquad r+3s = C^3$$

and therefore from $2r = (r-3s) + (r+3s)$ we have

$$(2A)^3 = B^3 + C^3$$

which is a solution of the original equation in strictly smaller integers. However, by applying the same argument to this new solution we can construct a strictly smaller solution, and so on, ad infinitum.

This is clearly impossible, since there must be some absolutely smallest integer solution. Consequently, by Fermat's principle of infinite descent, we see that solutions with z not divisible by 3 are impossible.

For the second case, suppose z is a multiple of 3. It follows that u is a multiple of 3, and v is not. In this case we cannot say that 2u is coprime to $u^2 + 3v^2$, because both are divisible by 3, but if we factor a 3 out of the quantity in parentheses in (1) we have

$$6u[\ 3(u/3)^2 + v^2\ ] = z^3 \quad (2)$$

so now 6u is coprime to the quantity in brackets, and so both factors are cubes, which implies

$$u = 36m^3 \qquad v^2 + 3(u/3)^2 = n^3$$

From Lemma 6 we have coprime integers r,s with s even, such that

$$u/3 = 3\ r^2\ s - 3\ s^3$$

which implies

$$u = 9s(r+s)(r-s) = 36m^3$$

so we have

$$4m^3 = s(r+s)(r-s)$$

and therefore

$$s = 4A^3 \qquad r+s = B^3 \qquad r-s = C^3$$

Since $2s + (r-s) = (r+s)$ we have

$$(2A)^3 + C^3 = B^3$$

so again we have a solution in smaller integers, and by the principle of infinite descent, this is impossible. Consequently, we have proven the result

THEOREM: The equation $x^3 + y^3 = z^3$ has no solution in non-zero integers.

7

Digit Reversal Sums Leading to Palindromes

Beginning with the decimal representation of any integer N, reverse the digits and add it to N. Iterate this operation. Typically you will soon arrive at a palindrome, i.e., a number that reads the same forwards and backwards. For example, starting with 39, we have 39 + 93 = 132. Then 132 + 231 = 363 = palindrome. Some numbers take a long time to yield a palindrome. For example, the sequence beginning with 89 is

```
        89      ———>    159487405
       187       |      664272356
       968       |     1317544822
      1837       |     3602001953
      9218       |     7193004016
     17347       |    13297007933
     91718       |    47267087164
    173437       |    93445163438
    907808       |   176881317877
   1716517       |   955594006548
   8872688       |  1801200002107
  17735476       |  8813200023188  = palindrome!
  85169247    ——>
```

Interestingly, there are twelve numbers less than 1000 for which the reverse-sum sequence leads to the palindrome

8813200023188, one of which, 484, is itself a palindrome. These are the longest finite sequences in this range.

It's worth noting that, if there were no carries in the addition, every number produced by adding a number to its reversal would be a palindrome. In fact, even with carries, it's easy to see that each digit of a number produced in this way can differ from the reflected digit by no more than 1. Compare, for example, the digits and reflected digits of the 4th from the last number in the sequence above

$$176881317877$$
$$778713188671$$
$$\overline{1010100010101}$$

This near palindromicity follows immediately from the fact that each digit and its reflection are both the sums of the same two digits, so the only way they can differ is if one involves a "carry" and the other doesn't. A carry is only a single unit, so the reflected digits can differ by (at most) only a unit from each other.

Despite the natural tendancy for the reverse-sum operation to produce palindromes, some sequences of reverse sums, such as the one beginning with the number 196, evidently NEVER yield a perfect palindrome. We say "evidently" because it has not been proven, but millions of terms of the "196 sequence" have been computed without ever reaching a palindrome. In fact, no one has ever proven that ANY number leads to an infinite sequence of palindrome-free numbers in the base 10.

On the other hand, it isn't hard to prove the existence of sequences that never produce a palindrome in certain other bases. For example, the smallest number that never becomes palindromic in the base 2 is 10110 (decimal 22). To prove this, first observe that the reverse-sum sequence beginning with 10110 is

$$10110$$
$$100011$$
$$1010100$$
$$1101001$$
$$10110100$$
$$\text{etc}$$

Digit Reversal Sums Leading to Palindromes

The last term quoted above is 10110100, which is of the form

$$10\ [n^*1]\ 01\ [n^*0]$$

where the symbol $[n^*x]$ signifies n consecutive occurences of the digit x. By simple arithmetic we can demonstrate that the reverse-sum sequence beginning with any number of this form proceedes as follows

$$10\ [n^*1]\ 01\ [n^*0]$$
$$11\ [(n-2)^*0]\ 1000\ [(n-2)^*1]\ 01$$
$$10\ [n^*1]\ 01\ [(n+1)^*0]$$
$$11\ [n^*0]\ 10\ [(n-1)^*1]\ 01$$
$$10\ [(n+1)^*1]\ 01\ [(n+1)^*0]$$

The last representation is identical to the first, except that n has been replaced by n+1. By induction, it follows that the entire sequence consists of repetitions of this cycle, and none of the elements are palindromes.

In the base 4, the number 255 (decimal) leads to a palindrome-free sequence with the following six-step cycle

$$22\ [n^*0]\ 131\ [n^*3]\ 12$$
$$10\ [(n+2)^*3]\ 23\ [(n+2)^*0]$$
$$11\ [n^*0]\ 3222\ [n^*3]\ 01$$
$$22\ [n^*0]\ 2111\ [n^*3]\ 12$$
$$10\ [(n+2)^*3]\ 23\ [(n+3)^*0]$$
$$11\ [(n+1)^*0]\ 312\ [(n+1)^*3]\ 01$$
$$22\ [(n+1)^*0]\ 131\ [(n+1)^*3]\ 12$$

A similar cycle exists for every base that is a power of two. In particular, if $B = 2^k$ then there is a cycle of length $2(k+1)$. For example, with the base $B = 8$ we have the eight-step cycle exemplified by the terms below

22000000000655577777777712   22 9*0 6555 9*7 12
44000000u004333377777777734
10777777777776700000000000
11000000000075677777777701
22000000000063577777777712

4400000000003737777777777734
10777777777777767000000000000
110000000000076667777777777701
2200000000006555777777777712   22 10*0 6555 10*7 12

Likewise for the base B = 16 = 2^4 we have the 2(4+1) = 10-step cycle exemplified by the terms shown below.

8800000008777ffffffff78   88 7*0 8777 7*f 78
10ffffffffef0000000000
1100000000fdeffffffff01
2200000000ebdffffffff12
4400000000c7bffffffff34
88000000007f7ffffffff78
10ffffffffef0000000000
1100000000feeeffffffff01
2200000000edddffffffff12
4400000000cbbbffffffff34
8800000008777ffffffff78   88 8*0 8777 8*f 78

The pattern for these powers of two is shown by taking representative terms from each cycle:

Base 2:   10 [n1] 01 [n0]
Base 4:   10 [n3] 23 [n0]
Base 8:   10 [n7] 67 [n0]
Base 16:   10 [nf] ef [n0]

and so on. In addition, there exist other self-similar cycles, beyond those in this infinite family. One example is this four-cycle in the base 2:

10 1111111111 0100000101111101 0000000000 00
11 0000000000 1000111011110001000 1111111111 01
10 1111111111 0100000101111101 00000000000 00
11 000000000000 1011111010000010 1111111111 01
10 1111111111111 0100000101111101 00000000000 00

Another example for the base 2 is shown below:

Digit Reversal Sums Leading to Palindromes

```
11 000000000 00011010 111111111 01
10 1111111111 01110011 000000000 00
11 000000000 100010000 111111111 01
10 1111111111 10010001 0000000000 00
11 0000000000 00011010 111111111 01
```

There are also a few sporadic examples in bases other than powers of 2, as shown by David Seal. However, no one knows how to construct a similar example for the base 10.

Empirically, the smallest numbers leading to palindrome-free sequences in each base from 2 through 19 are listed below (in decimal):

2	22	8	1021	14	361
3	100	9	593	15	447
4	255	10	196	16	413
5	708	11	1011	17	3297
6	1079	12	237	18	519
7	2656	13	2196	19	341

It's interesting that, in each base, all the palindrome-free sequences converge very rapidly on just a small number of sequences. For example, in the base 10 there are 63 numbers less than or equal to 4619 that (evidently) never become palindromic, and these 63 numbers each lead to one of only three palindrome-free sequences. The initial values of these sequences are

A	B	C
887	1857	9988
1675	9438	18887
7436	17787	97768
13783	96558	104547
52514	182127	930028
94039	903408	1750067
187088	1707717	9350638
1067869	8884788	17711177
etc	etc	etc

Could it be that these sequence are cyclical (in the sense of the base 2 and base 4 cycles described above), but with irrational periods? Notice that each term in the sequence can be regarded as a sort of "convolution" of the preceeding term, and there are known examples of sequences based on convolution that are cyclical with irrational periods. Some sequences in the base 3 seem to exhibit a degree of period near 13. Likewise in the base 10 there exist sequences with quasi-periodicity of order 8. In fact, sequence "C" in the table above shows this quasi-periodicity, beginning with 1750067.

8
Discordance Impedes Square Magic

As discussed in another note, the terms of a 3x3 magic square of squares can be re-arranged into a 3x3 bilinear array as shown below:

```
           |  2v  |  2v  |
     —   a^2    b^2    c^2
     2u
     —   d^2    e^2    f^2
     2u
     —   g^2    h^2    i^2
```

The values of 2u and 2v are the vertical and horizontal "partials", respectively. Notice that the differences are even, because we have the conditions

$$2e^2 = a^2 + i^2 = b^2 + h^2 = c^2 + g^2 = f^2 + d^2$$

which show that if any of the terms $a, b, ..., i$ is even, then they ALL must be even, so we can divide each of them by 2, and repeat until all are odd. (By a similar analysis we can show that all the terms $a, b, ..., i$ must be congruent modulo 3, so all the squares must be congruent modulo 36.) Therefore, we have

$$e^2 + 2u = b^2 \qquad e^2 - 2u = h^2 \qquad (1a,b)$$

$$e^2 + 2v = f^2 \qquad e^2 - 2v = d^2 \qquad (2a,b)$$

$$e^2 + 2(u+v) = c^2 \qquad e^2 - 2(u+v) = g^2 \qquad (3a,b)$$

$$e^2 + 2(u-v) = a^2 \qquad e^2 - 2(u-v) = i^2 \qquad (4a,b)$$

Subtracting equation (1b) from (1a) gives

$$u = \frac{b^2 - h^2}{4} = \left(\frac{b+h}{2}\right)\left(\frac{b-h}{2}\right)$$

Letting B and H denote the two integer factors in parentheses, we see that

$$u = BH \qquad B^2 + H^2 = e^2$$

Similarly from equations (2), (3), and (4) we deduce integers F, D, C, G, A, and I such that

$$v = FD \qquad F^2 + D^2 = e^2$$

$$u+v = CG \qquad C^2 + G^2 = e^2$$

$$u-v = AI \qquad A^2 + I^2 = e^2$$

These equations together imply that if a magic square of square exists then there must be four rational right triangles on a common diagonal with areas A1, A2, A3, A4 such that A1+A2 = A3 and A1-A2 = A4. This can also be expressed by the conditions

$$A3 + A4 = 2\,A1 \qquad A3 - A4 = 2\,A2$$

Just to emphasize how surprising a magic square of squares would be, it's worth noting that even if we drop the right hand requirement, it appears that there do not exist THREE rational right triangles on a common hypotenuse with areas in arithmetic

Discordance Impedes Square Magic

progression (signified by the left hand condition). This would correspond with a set of integers a,b,...,f,H such that

$$a^2 + b^2 = c^2 + d^2 = e^2 + f^2 = H^2$$
$$ab + cd = 2ef$$
(5)

No such integers have ever been found. However, if we relax the requirement by replacing H^2 with an arbitrary integer K, then there exist some solutions, although even these are not terribly plentiful. There are only 39 primitive solutions with K less than 52 million, as listed below:

K	diff	a	b	c	d	e	f
25345	5460	8	159	44	153	96	127
32266	6720	15	179	55	171	125	129
36490	6720	3	191	39	187	81	173
99025	15960	48	311	104	297	183	256
99125	9828	23	314	55	310	89	302
325117	61200	69	566	186	539	379	426
419050	87360	21	647	161	627	343	549
743665	143220	96	857	276	817	569	648
1006561	170940	81	1000	260	969	480	881
1229045	83160	439	1018	551	962	758	809
2047786	255360	5	1431	185	1419	375	1381
3129802	235200	661	1641	851	1551	1179	1319
4658425	228228	948	1939	1120	1845	1389	1652
5602945	840840	16	2367	376	2337	768	2239
7805890	779520	357	2771	651	2717	973	2619
8181625	879648	844	2733	1235	2580	1908	2131
9157850	993600	139	3023	473	2989	827	2911
9551777	1096200	604	3031	1001	2924	1484	2711
9699265	999180	672	3041	1036	2937	1473	2744
9887266	549120	1179	2915	1421	2805	1725	2629
14281930	2291520	33	3779	649	3723	1331	3537
14751841	3127320	79	3840	920	3729	2000	3279
18870865	3714480	96	4343	976	4233	2048	3831
19734650	1330560	1741	4087	2183	3869	2921	3347
21651370	2735040	891	4567	1551	4387	2389	3993
22163530	1404480	1629	4417	2023	4251	2513	3981
22884277	4395300	486	4759	1474	4551	2946	3769

K	diff	a	b	c	d	e	f
23386441	683760	1429	4620	1596	4565	1771	4500
26809445	4241160	778	5119	1679	4898	2911	4282
29713450	2735040	7	5451	511	5427	1029	5353
30002050	2542848	1917	5131	2555	4845	3669	4067
31676033	4781700	1057	5528	2023	5252	3488	4417
33826325	4823172	1223	5686	2185	5390	3698	4489
36133681	7332780	160	6009	1420	5841	3000	5209
37425389	5242860	1358	5965	2365	5642	4067	4570
37661026	2914560	1901	5835	2499	5605	3251	5205
38878705	4084080	1063	6144	1776	5977	2592	5671
42639466	7512960	875	6471	2135	6171	4021	5145
50205361	6846840	1280	6969	2360	6681	3769	6000

For a conjecture related to this table, see the note entitled No Four Rectangles in Line?

Suppose we place each of our three putative right triangles with their common hypotenuses on the x axis as shown below. (Only the a,b triangle edges are shown.)

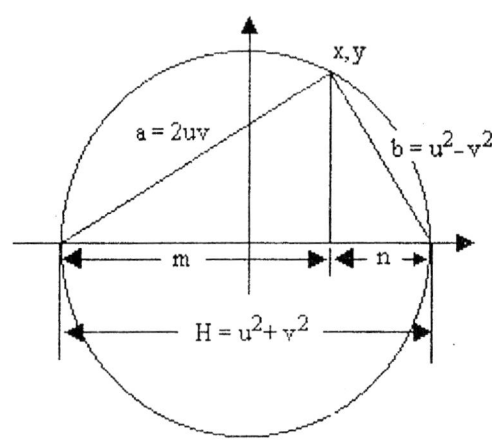

We want the areas of our three rational right triangles to be in arithmetic progression, and since the triangles are on a common base, it's clear that their heights must be in arithmetic progression.

Also, if the edges of the right triangle are rational, their areas and therefore their heights are rational, so the y coordinates of their upper vertices are rational. We can also show that the x coordinates of those vertices are rational because, referring to the above figure, we have

$$m = \frac{4 u^2 v^2}{(u^2 + v^2)} \qquad n = \frac{(u^2 - v^2)^2}{(u^2 + v^2)}$$

Therefore, if we re-scale the figure so that the radius of the circumscribing circle is unity (which we can do by dividing each length by half the length of the hypotenuse), the three upper vertices have rational x,y coordinates on the unit circle. Now, the usual way of parameterizing rational points on a unit circle is with the mapping

$$x = \frac{1 - t^2}{1 + t^2} \qquad y = \frac{2t}{1 + t^2} \qquad (6)$$

(It follows that $t = y/(1+x)$.) So, letting r,s,t denote the rational parameters of our three points on the circle, the requirement for their heights (i.e., their y coordinates) to be in arithmetic progression gives the following necessary condition for a complete solution of (5)

$$\frac{r}{1 + r^2} + \frac{s}{1 + s^2} = \frac{2t}{1 + t^2} \qquad (7)$$

It's important to note that although this is necessary, it is NOT sufficient, because while the rationality of the edges implies rationality of the x,y coordinates, the converse is not true. We can have rational solutions of (7) that don't correspond to triangles with rational edge lengths, even though the vertices have rational coordinates. Here are the first 41 primitive sets of three rational points on the unit circle with heights in arithmetic progression, which correspond to three rational right triangles on a common hypotenuse with areas in arithmetic progression:

Number Theory

H	del y	x1	y1	x2	y2	x3	y3	rs	r/s
65	-4	25	60	33	56	39	52	0.3333	1.3333
185	44	175	60	153	104	111	148	0.0833	0.3333
205	39	200	45	187	84	164	123	0.0370	0.3333
221	55	204	85	171	140	104	195	0.1200	0.3333
425	-104	119	408	297	304	375	200	0.1875	3.0000
725	161	696	203	627	364	500	525	0.0612	0.3333
901	-15	424	795	451	780	476	765	0.3333	1.0800
1105	-256	264	1073	744	817	952	561	0.2138	2.8739
1105	-276	1020	1020	817	744	1001	468	0.1067	2.1600
1469	185	1456	195	1419	380	1356	565	0.0133	0.3333
1769	-260	319	1740	969	1480	1281	1220	0.3333	2.0833
1885	-23	312	1859	427	1836	516	1813	0.6389	1.1206
1885	-231	464	1827	1003	1596	1300	1365	0.3333	1.8148
1885	-343	1643	1643	1365	1300	1624	957	0.1270	1.7076
2405	-68	483	2356	741	2288	925	2220	0.5439	1.2237
2425	-264	679	2328	1273	2064	1625	1800	0.3333	1.6875
2465	376	2431	408	2337	784	2175	1160	0.0208	0.3333
2465	729	2340	775	1953	1504	1044	2233	0.1026	0.2535
3445	-484	1749	2968	2387	2484	2805	2000	0.1829	1.7857
3485	-117	236	3477	925	3360	1276	3243	0.6365	1.3718
3625	-129	1276	3393	1577	3264	1820	3135	0.3986	1.2024
3965	649	3900	715	3723	1364	3416	2013	0.0248	0.3333
4505	-1036	689	4452	2937	3416	3825	2380	0.2449	3.0000
4745	976	4599	1168	4233	2144	3575	3120	0.0469	0.3333
4745	-329	2144	4233	2697	3904	3120	3575	0.2793	1.3518
5069	-455	1644	4795	2619	4340	3256	3885	0.3333	1.5306
5185	-243	144	5183	1575	4940	2196	4697	0.6189	1.5284
5525	511	5500	525	5427	1036	5304	1547	0.0068	0.3333
5525	-324	1131	5408	2163	5084	2805	4760	0.4643	1.4219
5525	961	5084	2163	4557	3124	3720	4085	0.0901	0.4614
6205	679	5244	3317	4747	3996	4080	4675	0.1317	0.6374
6565	-1551	1300	6435	4387	4884	5656	3333	0.2231	3.0000
6641	1420	6409	1740	5841	3160	4809	4580	0.0533	0.3333
6929	-920	1521	6760	3729	5840	4879	4920	0.7333	1.9200
7085	2759	7076	357	6363	3116	3960	5875	0.0134	0.0474
7565	1079	7476	1157	7227	2236	6800	3315	0.0178	0.3333
7565	-1344	2236	7227	4756	5883	6052	4539	0.2458	2.2121
8621	-2135	8155	8155	6171	6020	7696	3885	0.1157	2.0417
9425	-49	776	9393	1233	9344	1560	9295	0.7791	1.0882
9425	1587	8372	4329	7337	5916	5704	7503	0.1206	0.4905
9881	-2360	2169	9640	6681	7280	8569	4920	0.2133	3.0000

Discordance Impedes Square Magic

In none of these cases are the edge lengths of the triangles rational, so they don't represent solutions of (5). This raises the question of whether we can prove that a rational solution of (7) (which is a necessary condition for (5)) can NEVER give rational edge lengths.

If we could prove this, then it would follow that no solution of (5) exists, and therefore no magic square of squares exists. I don't have a proof, but it's interesting that we can rule out a large class of possible solutions by means of a certain "discordant form", as explained below.

Note that (7) is a quadratic in t, and if we solve for t we get an expression containing a square root of the quantity

$$[(1+r^2)(1+s^2)]^2 - [(r+s)(1+rs)]^2 \qquad (8a)$$

so a sufficient condition is for this to be a rational square. On this basis equation (8a) implies a rational number q such that

$$q^2 + [(r+s)(1+rs)]^2 = [(1+rs)^2 + (r-s)^2]^2$$

Dividing both sides by the right hand side, we can put this into a form that closely resembles the usual parametric representation of a circle

$$Q^2 + \left[\frac{\left(\frac{r+s}{1+rs}\right)^2}{1+\left(\frac{r-s}{1+rs}\right)^2} \right] = 1 \qquad (9)$$

where Q is an arbitrary rational number. One obvious class of solutions to (9), in view of (6), is given by identifying the quantity $(r-s)/(1+rs)$ with the rational parameter T, and then setting the quantity $(r+s)/(1+rs)$ to either $1-T^2$ or $2T$, equating the large fraction in the above expression with x or y (respectively) of equations (6).

In the first case, we set the quantity $(r+s)/(1+rs)$ to $1-T^2$, leading to a quadratic equation in r and s which can be solved for either variable to give an expression involving the square root of a number of the form $5q^2 + 2q + 5$ where q is rational, and we require this to be a rational square. Thus there must be integers m,n,k (not all divisible by 3) such that

$$5m^2 + 2mn + 5n^2 = k^2$$

but this is impossible, as can be seen by inspection modulo 9.

In the second case, after identifying $(r-s)/(1+rs)$ with the rational parameter T, we set quantity $(r+s)/(1+rs)$ to 2T, which implies the condition $r = 3s$, or, by symmetry, $r=s/3$ or $r=1/(3s)$ or $r=3/s$. This obviously gives an infinite family of solutions, i.e., sets of three rational points on a unit circle with heights in arithmetic progression, and in fact we see that 22 of the 41 sets listed above are of this form. In addition, three more are of the form $s=1/(3r)$, which is another infinite family of solutions, as shown in the note Bi-Rational Substitutions Giving Squares.

However, we can show that the EDGE LENGTHS corresponding to r,s are NEVER rational if $r = 3s$ or $1/(3s)$. To prove this, note that for any given vertex parameter t, corresponding to the vertex x,y, one of the edge lengths is given by

$$E^2 = (1-x)^2 + y^2$$

Substituting the expressions from (6) into this equation E^2 as a function of the parameter t

$$\frac{4t^2}{1 + t^2} = E^2$$

Consequently, for the two parameters r and s, with $r=3s$, we require both of the quantities $4s^2/(1+s^2)$ and $36s^2/(1+9s^2)$ to be rational squares. Inverting, and noting that the numerators are already squares, we see that we must have $1+s^2$

and $1+9s^2$ both equal to rational squares for some rational s. Clearing fractions, we need integers m,n such that

$$m^2 + n^2 = \text{square}$$

$$m^2 + 9n^2 = \text{square}$$

This is a problem of Concordant Forms, and it's known that there are no integers m,n that make both $m^2 + n^2$ and $m^2 + kn^2$ for many specific values of k, including 9. Thus there can be no solution to these equations. An almost identical argument works for the case $r=1/(3s)$.

Therefore, none of the rational solutions of (7) given by setting rs or r/s equal to 1/3 can possibly give three rational right triangles with areas in arithmetic progression. This shows an interesting relation between concordant forms and the obstacles to constructing a magic squares of squares.

9
Least Significant Non-Zero Digit of n!

Let p(k) be the least significant non-zero decimal digit of k! The first several values of this sequence are

2,6,4,2,2,4,2,8,8,8,6,8,2,8,8,6,8,2,4,4,8,4,6,4,4,8,4,6,8,...

Can we directly determine the kth term for any given k? Also, what is the asymptotic distribution of the digits? To answer these questions, let L(k) denote the least significant non-zero decimal digit of the integer k. Writing n! in the form

$$n! = (2^{a2})(5^{a5})(3^{a3})(7^{a7})...$$

we can let (n!)' denote this same number divided by its highest power of 10, i.e.,

$$(n!)' = (2^{(a2-a5)})(3^{a3})(7^{a7})...$$

Since we've divided out all powers of 10, the least significant digit of this number is non-zero, as are the least significant digits of the factors. Thus we have

$$L(n!) = L((n!)') = L[\ L(2^{a2} - a5)\ L(3^{a3})\ L(7^{a7})\ ...]$$

Least Significant Non-Zero Digit of n!

For any given integer n we can compute the exponent of any prime p in n! simply by summing the nearest integers to $[n/p^j]$ for j=1,2,..
For example, the exponent of 3 in 89! is given by

$$a_3 = [89/3] + [89/9] + [89/27] + [89/81]$$

$$= 29 + 9 + 3 + 1 = 42$$

Furthermore, the least significant decimal digit of 3^k is cyclical with the four values {1,3,9,7}, so it's easy to see that $L(3^{42}) = 9$.

Likewise, the least significant digits of the sequence p^k, k=1,2,.. for every odd prime ending with the digit 3 or 7 has a period of four, while those ending with 9 have a period of two, and those ending with 1 have a period of one. Thus, all these periods are divisors of four.

Of course, the least significant digits of 2^k also have a period of four, i.e., {2,4,8,6}.

Also, as we multiply successive integers to generate n!, we always have more powers of 2 than of 5, so the value of L(n!) is easily computed recursively as L(L(n)L((n-1)!)) unless n is a multiple of 5, in which case we need more information. As a result, the values of L(n!) come in fixed strings of five, as shown below

				n				
1	5	10	15	20	25	30	35	40	45

L(n!): 1264 22428 88682 88682 44846 44846 88682 22428 22428 66264 ...

Thus, if we know the value of L((5n)!) we automatically know the values of L((5n+j)!) for j=0,1,2,3,4 However, the pattern of the values L((5n)!) is not immediately apparent. If we tabulate these values we find that they too come in fixed strings of five, so we only need to know L((25n)!) to automatically know L((25n+5j)!) for j=0,1,2,3,4.

Continuing in this way, we can tabulate the values of $L((5^t n)!)$ as shown below

				n				
1	5	10	15	20	25	30	35	40	45

```
L(    n!): 1264 22428 88682 88682 44846 44846 88682 22428 22428 66264
L(   5n!): 2884 48226 24668 48226 48226 86442 24668 62884 24668 24668
L(  25n!): 4244 82622 82622 28488 46866 64244 82622 82622 28488 46866
L( 125n!): 8824 68824 26648 68824 42286 26648 26648 42286 26648 84462
L( 625n!): 6264 22428 88682 88682 44846 44846 88682 22428 22428 66264
            etc.
```

Notice that the pattern of digits for L(625n!) is the same as for L(n!). In general it appears that the pattern for $L((5^j n)!)$ is the same as for $L((5^{(j+4)} n)!)$. In addition, on each level there are precisely four distinct blocks of 5 sequential digits, one block beginning with each of the digits 0,2,4,6,8.

From the above tabulations we can extract the essential patterns for $L((5^k n)!)$

Look-Up Table for L() Patterns

k mod 4					
0	06264	22428	44846	66264	88682
1	02884	24668	48226	62884	86442
2	04244	28488	46866	64244	82622
3	08824	26648	42286	68824	84462

This table represents all we need to determine the value of L(n!) for any integer n. First we convert n to the base 5, so we have

$$n = d\_0 + d\_1 * 5 + d\_2 * 5^2 + \ldots + d\_h * 5^h$$

Now we enter the above table at the row h (mod 4) in the block whose first digit is 0 (because the coefficient of $5^{(h+1)}$ is zero), and determine the digit in the (d_h)th position of this block. Let this digit be denoted by s_h. Then we enter the table at row h-1 (mod 4) in the block that begins with s_h, and determine the digit in the (d_(h-1))th position of this block. Let

Least Significant Non-Zero Digit of n! 73

this digit be denote by $s_{(h-1)}$. We continue in this way down to s_0, which is the least significant non-zero digit of n!. To illustrate, consider the case of the decimal number n=1592. In the base 5 this is n=22332. Now we enter the above table at row k=4=0 (mod 4) in the block beginning with 0, which is 06264. The leading digit of n (in the base 5) is 2, so we check the digit in position 2 of this block to give $L((2*5^4)!) = 2$. Then we enter the table at row k=3 (mod 4) in the block beginning with 2, which is 26648, to find $L((2*5^4 + 2*5^3)!) = 6$.

Then in the row k=2 (mod 4), the block beginning with 6 is 64244, and we find $L((2*5^4 + 2*5^3 + 3*5^2)!) = 4$. From this we know we're in the block 48226 in row k=1 (mod 4), so we have $L((2*5^4 + 2*5^3 + 3*5^2 + 3*5)!) = 2$. Finally, we enter the row k=0 (mod 4) in block 22428 to find the result

$$L(1592!) = L((2*5^4 + 2*5^3 + 3*5^2 + 3*5 + 2)!) = 4$$

To streamline this process, let's define an array A(4,5,5) where the first index signifies the row (0,1,2,3), the second is the block selector (0,2,4,6,8) in that row, and the third is the digit number (0,1,2,3,4) in that block. If it's understood that the first index is to be taken modulo 4, and if we let dj denote the jth digit of the base-5 representation of n, then the above evaluation be written in the form

A(4, 0, d4) = s4
A(3, s4, d3) = s3
A(2, s3, d2) = s2
A(1, s2, d1) = s1
A(0, s1, d0) = s0 = $L((d4*5^4 + d3*5^3 + d2*5^2 + d1*5 + d0)!)$

This shows how we can easily determine the value of L(n!) by means of k look-ups (in a simple fixed 4x5x5 table) where k is the number of base-5 digits of n. From this we can also rigorously determine the distribution of digits, which the table's symmetry seems to imply must be uniform. Just checking empirically, we find the following distribution of the values of L(n!) for n from 2 to 10^t with t=4,5,6. (This excludes n=1 for which L(n!)=1.)

	2	4	6	8
10^4	2509	2486	2494	2510
10^5	25026	24999	24973	25001
10^6	249993	250013	250040	249953

Naturally a similar analysis can be performed with respect to the least significant digit of n! in any other base. For example, in the base 3, we find that the blocks on the even levels are 112 and 221, and the blocks on the odd levels are 122 and 211. With this information we can construct the function table shown below.

previous output	current digit	output
1	0	1
1	1	p+1
1	2	2
2	0	2
2	1	2-p
2	2	1

where "p" denotes the parity of the exponent of 3 for the current digit. To illustrate, suppose we wish to determine the least significant non-zero base-3 digit of (139!). The number 139 written in the base 3 is 12011, so the exponent of 3 for the leading digit is 4, which has parity 0. Thus the parity string of the exponents is 01010. Beginning with the most significant digit, 1, and a "previous output" of 1 (which is always the initial "previous output") we enter the table in the row 1 1 to find that the output is p+1, which equals 1 (because the current exponent parity is p=0). Then we take this output and the next input digit, 2, and enter the table at the row 1 2 to get the output 2. Then we take this output and the next input digit, 0, and enter the table in the row 2 0 to find the output 2. Next we enter at 2 1 to find he output 2-p, and on this level we have p=1, so the output is 1. Finally we enter the table at row 1 1 to find the output p+1, and on this level we have p=0, so the final output is 1.

Least Significant Non-Zero Digit of n!

This process essentially acts as a kind of "filter", taking consecutive digits from the set {0,1,2} and outputting digits from the set {1,2}, just as the decimal algorithm takes a stream of digits from the set {0,1,...,9} and outputs a stream of digits from the set {2,4,6,8}.

For the base-3 filter, a continuous stream of "0" input digits will leave the output unchanged, i.e., it will retain the previous output value. On the other hand, a continuous stream of "2" input digits will cause the output to oscillate between 1 and 2 on each step. A continuous stream of "1" input digits will act like "0" when the parity is even, and will act like "1" when parity is odd, with the result being that the output will change state on the odd steps.

10

Geodesic Diophantine Boxes

Do there exist rectangular solids with integer edge lengths such that all three geodesic distances between opposite vertices are also integers? Herman Baer noted the three solutions (108,357,368), (564,748,1425), and (348,975,2380), and asked for more information about such solutions. This is an interesting problem - one that you'd think would have been considered before - but I haven't been able to find any reference to this particular set of Diophantine equations.

Dave Rusin responded to Baer with some additional solutions, such as (243984,675500,689613). I was curious to know how many primitive solutions exist between the three small examples and this larger one.

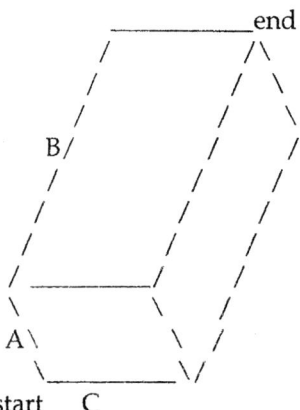

Geodesic Diophantine Boxes

To review, the problem is to find a rectangular box with integer dimensions AxBxC such that all the (locally) geodesic paths on the surface of the box from one corner to the diagonally opposite corner, as shown below, have integer lengths.

Clearly any geodesic path on the surface must be a straight line on any given wall, so if we "unfold" the box in various ways and lay all the sides out flat, the geodesic paths are just the straight lines connecting opposite corners, as illustrated below (where "s" indicates the starting corner and e1,e2,e3 are the three basic positions of the ending corner for the three different ways of unfolding the box.

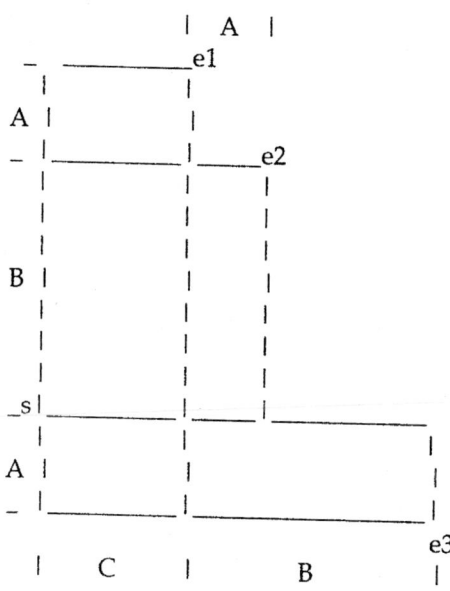

From this we see that the three locally geodesic path lengths x,y,z are related to the edge lengths A,B,C according to the equations

$$x^2 = (A+B)^2 + C^2$$

$$y^2 = (A+C)^2 + B^2$$

$$z^2 = (B+C)^2 + A^2$$

So, the task is to find integers A,B,C,x,y,z that satisfy all three of these equations simultaneously. If we re-write these equations in the form

$$2ab + D2 = x^2 \qquad 2ac + D2 = y^2 \qquad 2bc + D2 = z^2$$

where D2 is the squared length of the main diagonal of the box (which we don't require to be a square integer), this Diophantine problem bears some resemblence to a class of problems considered by Diophantus himself, namely, to find three numbers such that their products in pairs increased by a constant are each squares. In other words, Diophantus considered the set of simultaneous equations (in rational numbers)

$$ab + k = x^2 \qquad ac + k = y^2 \qquad bc + k = z^2$$

The case k=1 in particular has been studied extensively, and extended to sets of four, and even five (rational) numbers whose products in pairs are 1 less than squares. This was studied by Euler and many others. However, the Diophantine box problem is a bit trickier, because it asks us to find three integers such that TWICE each pair-wise product, increased by the sum of of the squares, is a square.

I'm not sure if the classical methods used to solve Diophantus' original system of equations can be adapted to treat this problem, but we can say a few things about the integer $S = A+B+C$ that makes searching for solutions a bit easier. Obviously each of the conditions corresponds to a Pythagorean triple which can be parameterized in the usual way, i.e.,

$$(ku^2 - kv^2)^2 + (2kuv)^2 = (ku^2 + kv^2)^2$$

where u,v are coprime (one odd, one even). Now, $S = A+B+C$ is the sum of the two terms on the left, so essentially the conditions of the problem require that S for a solution be expressible in the form

$$(u^2 - v^2) + (2uv)$$

in (at least) three distinct ways. Equivalently, S must have at least three distinct representations of the form $(u+v)^2 - 2v^2$. Thus, we are dealing with the number field $Q[\sqrt{2}]$, which possesses unique factorization. By Fermat's "Christmas theorem" we know that S must be a product of at least two primes of the form $X^2 - 2Y^2$, and probably three. Also we can see that a primitive solution must not be divisible by any primes NOT of this form.

At first it may seem surprising that a value of S having only 2 factors could yield a solution, because S must be expressible in THREE distinct ways as a sum of two legs of a Pythagorean triple, viz, in the quadratic form $X^2 - 2Y^2$. Thus, we might expect that S must have at least three prime divisors congruent to +1 or -1 (mod 8). However, there exist solutions such as S = (41)(17921) in which S does have just two prime factors. This is quite rare - the only other such solution less than 10 million is S = (89)(88001). The reason such solutions are possible is that not all of the Pythagorean triples are necessarily primitive, and, furthermore, the common divisor "k" for any non-primitive triples must be a divisor of S=pq. As a result, we can get two of our representations of S from the usual composition of the quadratic forms of the two prime factors p,q of S, and then we get the 3rd representation as p times the minimal representation of q.

In any case, the acceptable S values must be composed entirely of at least two (and usually three) primes, not necessarily distinct, modulo which 2 is a quadratic residue, i.e., primes congruent to +-1 modulo 8. The first few primes of this type are

7   17   23   31   41   47   71   73   79   89   97   103...
etc.

Indeed, the three smallest solutions have the S values

$$833 = (7)(7)(17) \qquad 2737 = (7)(17)(23) \qquad 3703 = (7)(23)(23)$$

In general, to find values of S that give primitive solutions we can examine each integer and immediately exclude any that are not composed entirely of 2 or more primes congruent to +- 1 (mod 8).

Now, if two or all three of the Pythagorean triples are primitive, we can infer the solution simply by finding two minimal representations of S in the form $2y^2 - x^2$ where $0 < x < y$, which can clearly be found by a 1-dimensional search up to sqrt(S). Once we have found these minimal representations, we can take them, two at a time, to see if they give a valid solution.

For example, the minimal representations of 833 in the form $2y^2 - x^2$ are $(x,y) = (7,21)$, $(15,23)$, and $(25,27)$, and these are converted to (u,v) parameterizations of the respective Pythagorean triples by

$$u = y \qquad v = y - x$$

Taking the 2nd and 3rd minimal forms gives the triples

u	v	$u^2 - v^2$	$2uv$
23	8	465	368
27	2	725	108

Recalling that our three triples are of the form $(A+B,C)$, $(A+C,B)$, and $(B+C,A)$, it's clear that if this is a solution then we must have A=108, B=368, and C = 465-108 = 725-368 = 357, and we can easily verify that this is, in fact, a solution.

On the other hand, if our solution involves two (or more) non-primitive Pythagorean triples, the above method won't work, because the primitive triples are based on the minimal forms of the entire S, whereas the non-primitive triples are based on the minimal forms of proper FACTORS of S. So, to find such solutions we need to proceed as before, but now we need to include, in addition to the minimal representations of S, the representations of the form $n(2y^2 - x^2)$ where n is any proper divisor of S. For any divisor n we find the minimal form(s) of S/n, convert them to (u,v) paramaters as above, and then the resulting Pythagorean legs are $n(u^2 - v^2)$ and 2nuv. We can then check these purported triples in pairs as above to determine if they give three suitable triples.

Obviously if (A,B,C) is a solution then so is (kA,kB,kC) for any integer k, so we've considered only the "primitive" solutions, i.e., those for which gcd(A,B,C) = 1. Now, each primitive solution leads to the three Pythagorean triples

Geodesic Diophantine Boxes

$$(A+B)^2 + C^2 = x^2$$
$$(A+C)^2 + B^2 = y^2$$
$$(B+C)^2 + A^2 = z^2$$

where x,y,z are the three geodesic path lengths. Notice in the attached table that almost all the solutions less than 3 million have two primitive Pythagorean triples. For example, the smallest solution is

S	A	B	C	(prime factors of S)		
	A+B	C	x	k1	u1	v1
	A+C	B	y	k2	u2	v2
	B+C	A	z	k3	u3	v3
= 833	108	357	368	(7)(7)(17)		
	465	368	593	1	23	8
	476	357	595	119	2	1
	725	108	733	1	27	2

and we note that two of the three Pythagorean triples are primitive, but one is just the simple $3^2 + 4^2 = 5^2$ triple multiplied by 119. (Oddly enough, the larger solution found by Rusin by means of elliptic curves also involved a multiple of the $3^2 + 4^2 = 5^2$ triple.) Most of the solutions are of this type, i.e., two primitive triples and one non-primitive triple, but there are also solutions in which only one of the Pythagorean triples is primitive, such as

1228687	112812	431955	683920	(79)(103)(151)		
	544767	683920	874367	103	83	40
	796732	431955	906293	1	818	487
	1115875	112812	1121563	79	119	6

Furthermore, there are even primitive geodesic boxes for which NONE of the Pythagorean triples is primitive, including

2627639	337824	516260	1773555	(7)(17)(71)(311)		
	854084	1773555	1968491	7	517	118
	2111379	516260	2173579	311	83	10
	2289815	337824	2314601	17	368	27

In addition, the table also shows that we have some "ultra-primitive" geodesic boxes, i.e., primitive geodesic boxes such that all three Pythagorean triples are also primitive. The smallest example of this is

```
2107609  312040  587340 1208229      (7)(17)(89)(199)
         899380 1208229 1506221         1  1165  386
        1520269  587340 1629781         1  1255  234
        1795569  312040 1822481         1  1345  116
```

which has the dimensions A=312040, B=587340, C=1208229. The geodesic path lengths for this box are given by three primitive Pythagorean triples as

$$x = (1069)(1409) \qquad y = (197)(8273) \qquad z = (1822481)$$

and the sum of the edge lengths is

$$S = A+B+C = 2107609 = (7)(17)(89)(199)$$

There are a total of 8 ultra-primitive geodesic boxes with S less than 50 million, as shown in the table.

Having found three-dimensional boxes of size AxBxC such that the edge lengths and all three geodesic path lengths connecting opposite vertices are integers, it's interesting to consider the analagous question for 4-dimensional boxes. First, note that for a 2D box of size AxB there is essentially just one geodesic distance S along the surface connecting opposite corners, given by

$$S^2 = (A+B)^2$$

In other words, the distance is simply A+B. For a 3D box of size AxBxC there are three local geodesic distances, given by

$$S1^2 = (A+B)^2 + C^2$$
$$S2^2 = (A+C)^2 + B^2$$
$$S3^2 = (B+C)^2 + A^2$$

The usual way of seeing this is to simply unfold the box and draw the straight lines connecting opposite corners, be we

Geodesic Diophantine Boxes

could also deduce this more formally by noting that we're required to go from the point [0,0,0] to the point [A,B,C] in the minimum linear segments, each of which is on (at least) one surface of the box. In other words, every segment must have at least one of the components at its min or max limit (0 or A,B,C respectively) for the entire segment. So the shortest path must be of the form

$$[0,0,0] \to [A,0,t] \to [A,B,C]$$

Notice that for the first segment the 2nd coordinate is always 0, whereas for the second segment the 1st coordinate is constantly A, so both segments are on the surface. The total length of this path

$$S = \sqrt{A^2 + t^2} + \sqrt{B^2 + (C-t)^2}$$

Differentiating with respect to t and setting the result to zero, we find that $t = AC/(A+B)$, which can be substituted back into the above equation to give the expected result

$$S^2 = (A+B)^2 + C^2$$

Now, for a 4-dimensional box we need to go from [0,0,0,0] to [A,B,C,D] via the most economical path on the surface. In this case the "surface" consists of 8 solid regions in which three of the coordinates are anywhere between their min and max while the fourth coordinate is at its min or max. Again we can visualize this by "unwrapping" the box, placing two of the faces adjacent to each other. For example, if we unfold a face with dimensions AxCxD into the same 3D space as a face of dimensions BxCxD we have a solid with overall dimensions (A+B)xCxD, and the shortest path connecting the two opposite corners has length S given by

$$S^2 = (A+B)^2 + C^2 + D^2$$

Likewise for the other face combinations. Alternatively, we can proceed more formally by considering the path

$$[0,0,0,0] \quad \rightarrow \quad [A,0,u,v] \quad \rightarrow \quad [A,B,C,D]$$

whose total length is

$$S = \text{sqrt}[A^2 + u^2 + v^2] + \text{sqrt}[B^2 + (C-u)^2 + (D-v)^2]$$

Here we have two unknown paramaters, so we differentiate with respect to each of them separately to give the two conditions

$$dS/du = 0 \qquad dS/dv = 0$$

Solving these simultaneous equations for u and v gives

$$u = AC/(A+B) \qquad v = AD/(A+B)$$

Substituting these back into the preceeding expression gives the expected result $S^2 = (A+B)^2 + C^2 + D^2$. In general, if the edge lengths of an n-dimensional rectangular box are E1, E2,..,En then the lengths of the geodesics connecting opposite vertices on the surface are of the form

$$S^2 = (E1+E2)^2 + E3^2 + E4^2 + \ldots + En^2$$

for all $n(n-1)/2$ distinct permutations of the edges. Thus, for the case of a 4-dimensional box we have six distinct geodesic lengths

$$
\begin{aligned}
S1^2 &= (A+B)^2 + C^2 + D^2 \\
S2^2 &= (A+C)^2 + B^2 + D^2 \\
S3^2 &= (A+D)^2 + B^2 + C^2 \\
S4^2 &= (B+C)^2 + A^2 + D^2 \\
S5^2 &= (B+D)^2 + A^2 + C^2 \\
S6^2 &= (C+D)^2 + A^2 + B^2
\end{aligned}
$$

So, the question is whether there exist 4-tuples of integers A,B,C,D such that all six of these lengths are integers. It's interesting that this can be expressed in terms very similar to the ancient Diophantine problem of finding four numbers whose six pairwise products increased by a constant n are all squares.

Geodesic Diophantine Boxes

However, in this case we need to consider TWICE the pairwise products, because we have

$$S1^2 = 2AB + K$$
$$S2^2 = 2AC + K$$
$$S3^2 = 2AD + K$$
$$S4^2 = 2BC + K$$
$$S5^2 = 2BD + K$$
$$S6^2 = 2CD + K$$

where $K = A^2 + B^2 + C^2 + D^2$ is the squared main diagonal. The problem involving just the pairwise products (without the factor of 2) was treated by Euler, but unfortunately the method he used is not applicable when the factor of 2 is included (because it relies on being able to set $4 - y^2$ to zero with a rational number y, but this term becomes $8 - y^2$ when the factor of 2 is included).

I haven't found any non-trivial integer solutions to this set of equations. Does anyone know of any solutions? Or a proof that such solutions are impossible?

By the way, just for fun, we might consider the general ways in which a sum of squares can be defined on a 4-part partition of a given number $S = A+B+C+D$. We have the following possibilities

Form	Number of distinct values under permutations of A,B,C,D
$(A)^2 + (B)^2 + (C)^2 + (D)^2$	1
$(A+B)^2 + (C)^2 + (D)^2$	6
$(A+B+C)^2 + (D)^2$	4
$(A+B)^2 + (C+D)^2$	3
$(A+B+C+D)^2$	1

Since every number is expressible as a sum of four squares, it's clear that the first form, which is invariant under permutations, can be set equal to a square. Also, the last form is invariant and is automatically a square. In addition, it's not hard to find sets of integers [A,B,C,D] such that all 4 of the permutations of the 3rd form are squares, as shown by [189,1032,564,952], which gives

$$(189+1032+564)^2 + (952)^2 = (2023)^2$$
$$(189+1032+952)^2 + (564)^2 = (2245)^2$$
$$(189+564+952)^2 + (1032)^2 = (1993)^2$$
$$(1032+564+952)^2 + (189)^2 = (2555)^2$$

(Notice that this solution has S = 189+1032+564+952 = 2737, which is the 2nd smallest S value for solutions of the 3D box problem.)

In addition, we can find integers A,B,C,D such that all three of the permutations of the 4th form are squares, as shown by [23,47,55,113], which gives

$$(23+47)^2 + (55+113)^2 = (182)^2$$
$$(23+55)^2 + (47+113)^2 = (178)^2$$
$$(23+113)^2 + (47+55)^2 = (170)^2$$

So it's easy to find solutions for all of the forms EXCEPT the one that gives the geodesic lengths of a 4D box, which is clearly the most stringent because it imposes SIX Diophantine conditions on four numbers.

Regarding the 3D case, all the solutions for S=A+B+C less than 50 million were given above, including eight "ultra-primitive" solutions.

Only one solution for each S is listed, and I think in most cases it is the ONLY solution. The only exception to this for S less than 1 million is S=82943, which actually has two solutions

$$82943 = 6647 + 33228 + 43068$$
$$= 13940 + 31155 + 37848$$

Multiple representations of other forms seem to be more common. For example, the number 1666 can be partitioned into four parts A,B,C,D in 15 distinct ways such that $(A+B)^2 + (C+D)^2$ is a square for all the permutations of those four parts (although only one of them, 137 353+577+599, involves just prime numbers).

11

Highly Heronian Ellipses

A triangle with integer edge lengths and integer area is called "Heronian", after the Greek mathematician Heron. Such triangles are also sometimes called "rational triangles", since any triangle with rational edges and area can obviously be scaled to give one with integer edges and area. Of course, every ordinary Pythagorean triple (a,b,c) gives a rational triangle, since the area of this right triangle is simply ab/2. (For example, the area of the familiar 3,4,5 triangle is 6.) So, to make things interesting, we usually focus on triangles that are not right-angled.

Heron noted the example (13,14,15), which has an area of 84. This can easily be computed using Heron's formula for the area of a general triangle as a function of it's three edge lengths a,b,c:

$$A = \frac{1}{4} \sqrt{(a+b+c)(-a+b+c)(a-b+c)(a+b-c)}$$

It s not hard to find infinite families of such triangles.

For example, in 598 AD Brahmagupta noted that for any integers a,b,c we have an oblique triangle with edges of length

$$L1 = (a^2 + b^2) c$$
$$L2 = (a^2 + c^2) b$$
$$L3 = (a^2 - bc)(b + c)$$

whose area is $A = abc(b+c)(bc-a^2)$. Gauss (as usual) went further and found that ALL Heronian triangles have edge lengths that can essentially be expressed in the form

$$L1 = 4abfg[a^2 + b^2]$$
$$L2 = +-4ab(f+g)(fa^2 - gb^2)$$
$$L3 = 4ab[(af)^2 + (bg)^2]$$

in which case the radius of the circumscribing circle is

$$R = (a^2 + b^2)[(af)^2 + (bg)^2]$$

Recall that the area of a triangle is $L1*L2*L3/(4R)$ where R is the circumscribing radius. Hence the area of Gauss' Heronian triangles is

$$A = |16(ab)^3 \, fg(f+g)(fa^2 - gb^2)|$$

It may be worth mentioning that if we define

$$L1 = x + y \qquad L2 = x + z \qquad L3 = y + z$$

then from Heron's area formula the triangle is Heronian if an only if $xyz(x+y+z)$ is a square.

We can also find specialized Heronian triangles. For example, in 1722 the Japanese mathematician Nakane Genkei gave a recurrence for an infinite family of triangles with *consecutive* integer sides and integer areas (although it isn't clear if he completed the induction).

We'll derive Genkei's result below, but first let's consider a slightly more general problem, namely, to find the Heronian triangles whose edges a,b,c are in arithmetic progression, i.e., for some integers x and d we have the edge lengths

$$a = x - d \qquad b = x \qquad c = x + d$$

Notice that if x and d have a greatest common factor g, then g divides each of a, b, and c, which implies that g appears to the 4th power inside the radical sign in Heron's formula, so it can be removed from the square root. Thus, g^2 must divide A, so we need only consider primitive solutions, i.e., those with

$g=1$. In other words, we can assume that x and d have no common factor. Substituting into Heron's formula gives

$$4A = \sqrt{(3x)(x+2d)(x)(x-2d)}$$

Squaring both sides, we have

$$(4A)^2 = 3x^2 [(x+2d)(x-2d)]$$

Since the left side is even, we know that x must be even, so there is an integer y such that $x=2y$. Also, by unique factorization of integers, it's clear from this equation that the quantity in [] brackets must be 3 times a square, which means there is an integer m such that

$$(x+2d)(x-2d) = 3m^2$$

Since x and d have no common factor, it's clear that the two factors on the left are coprime except for a possible common factor of 2 because $x=2y$. It follows that m must also be even, so we have an integer n such that $m=2n$. With these substitutions we have

$$y^2 - d^2 = 3n^2$$

in coprime integers y,d,n, which can be written as

$$d^2 + 3n^2 = y^2$$

The general solution of this Diophantine equation is

$$|d| = \frac{p^2 - 3q^2}{2} \qquad |n| = pq \qquad |y| = \frac{p^2 + 3q^2}{2}$$

for odd integers p,q, or double these quantities if p and q have opposite parity. Therefore, with odd p,q we have

$$|x| = (p^2 + 3q^2) \qquad |d| = (p^2 - 3q^2)/2$$

For example, with p=q=1 this gives x=4,d=1, leading to the Pythagorean triangle with edge lengths a=3,b=4,c=5. On the other hand, if p,q, have opposite parity we have

$$|x| = 2(p^2 + 3q^2) \qquad |d| = (p^2 - 3q^2)$$

The first example of this type is with p=2,q=1, which gives x=14, d=1, leading to the original Heronian example a=13,b=14, c=15. In general, the case with p,q both odd gives

$$a = (p^2 + 9q^2)/2$$
$$b = p^2 + 3q^2 \qquad A = 3pq(p^2 + 3q^2)/2$$
$$c = 3(p^2 + q^2)/2$$

whereas if p,q have opposite parity we have

$$a = p^2 + 9q^2$$
$$b = 2p^2 + 6q^2 \qquad A = 6pq(p^2 + 3q^2)$$
$$c = 3p^2 + 3q^2$$

Here's a table of all the Heronian triangles with edges in arithmetic progression and total perimeter less than 1000 given by odd values of p and q:

p	q	perimeter	a	b	c	Area
1	1	12	5	4	3	6
1	3	84	41	28	15	126
5	-1	84	17	28	39	210
5	3	156	53	52	51	1170
7	1	156	29	52	75	546
-1	5	228	113	76	39	570
7	-3	228	65	76	87	2394
7	5	372	137	124	111	6510
11	1	372	65	124	183	2046
1	7	444	221	148	75	1554
11	-3	444	101	148	195	7326
5	7	516	233	172	111	9030
13	-1	516	89	172	255	3354
11	5	588	173	196	219	16170

Highly Heronian Ellipses

p	q	perimeter	a	b	c	Area
13	3	588	125	196	267	11466
13	-5	732	197	244	291	23790
-1	9	732	365	244	123	3294
-5	9	804	377	268	159	18090
11	-7	804	281	268	255	30954
7	9	876	389	292	195	27594
17	-1	876	149	292	435	7446
13	7	948	305	316	327	43134
17	3	948	185	316	447	24174

Notice that, with the exception of the case p=1,q=1, all the perimeters appear twice. In each case we have (with an appropriate choice of signs) p + q equal to the same quantity, and the edge length b is the same for both. These two solutions correspond to the two roots of the combined equations $p^2 + 3q^2 = b$ and p + q = n for fixed values of b and n.

The next solution set with odd p,q has four cases, with a perimeter of 1092 and b = 364:

p	q	perimeter	a	b	c	Area
1	11	1092	545	364	183	6006
17	-5	1092	257	364	471	46410
11	9	1092	425	364	303	54054
19	1	1092	185	364	543	10374

This is due to the fact that 1092 is expressible in the form $3(p^2 + 3q^2)$ in two distinct ways

$$1092 = 3[(19)^2 + 3(1)^2] = 3[(17)^2 + 3(5)^2]$$

which is due to the fact that it's divisible by two primes, 7 and 13, congruent to 1 mod 3. These two expressions arise from the different ways of multiplying out the product

$$1092/3 = 364 = (2)^2 (91) = 4[7][13]$$

where

$$7 = (2)^2 + 3(1)^2 \qquad 13 = (1)^2 + 3(2)^2$$

92 Number Theory

using the ancient product rule for quadratic forms

$$(x^2 + N y^2)(u^2 + N v^2) = (xu + Nyv)^2 + N(xv - yu)^2$$
$$= (xu - Nyv)^2 + N(xv + yu)^2$$

Thus, each distinct prime factor of the form 3k+1 doubles the number of distinct solutions (in general). If we divide each of the perimeters in the above table by 3(4) we find

7 13 19 31 37 43 7^2 61 ...

which are precisely the primes congruent to 1 mod 3 (or squares of those primes), leading to the two solutions for each perimeter. When we reach perimeters, like 1092, that are (12 times) the product of TWO distinct primes congruent to 1 mod 3, we have $2^2 = 4$ distinct solutions. Likewise when we reach the perimeter 12(7*13*19) = 20748 with THREE distinct prime factors congruent to 1 mod 3 we expect to find $2^3 = 8$ distinct Heronion triangles with that perimeter, and in fact we have

p	q	perimeter	a	b	c	Area
17	47	20748	10085	6916	3747	8288826
79	-15	20748	4133	6916	9699	12293190
83	-3	20748	3485	6916	10347	2583126
37	43	20748	9005	6916	4827	16505034
-29	45	20748	9533	6916	4299	13538070
53	-37	20748	7565	6916	6267	20343414
71	25	20748	5333	6916	8499	18413850
73	23	20748	5045	6916	8787	17417946

A geometrical way of looking at this is to imagine a planet in an elliptical orbit whose eccentricity is 1/2 and such that the distance between the two foci of the orbit is 6919 units. If we trace out the locus of points relative to that fixed line segment such that the sum of the distances from the two foci is constant, we have an ellipse. Thus we can imagine a planet following an elliptical path such that the perimeter of the triangle formed by the planet and the two foci remains constant at 20748 units.

Highly Heronian Ellipses

The above solutions are then seen to determine the locations of 32 "Heronian points" on that ellipse (because each of the 8 solutions appears in each of the four quadrants) such that the distances to the two foci are both integers AND the area of the triangle formed by the foci and that point is an integer. This is illustrated in the figure below:

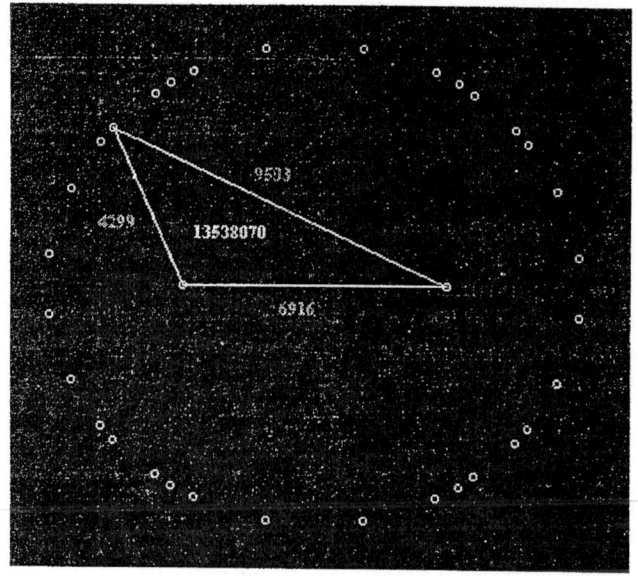

Moving on to the solutions given by values of p,q with opposite parity, we have the following table of solutions for perimeters less than 1000:

p	q	perimeter	a	b	c	Area
2	1	42	13	14	15	84
1	2	78	37	26	15	156
4	1	114	25	38	51	456
2	3	186	85	62	39	1116
5	2	222	61	74	87	2220

4	3	258	97	86	75	3096
1	4	294	145	98	51	1176
7	2	366	85	122	159	5124
8	1	402	73	134	195	3216
5	4	438	169	146	123	8760
2	5	474	229	158	87	4740
8	3	546	145	182	219	13104
4	5	546	241	182	123	10920
7	4	582	193	194	195	16296
10	1	618	109	206	303	6180
1	6	654	325	218	111	3924
10	3	762	181	254	327	22860
11	2	798	157	266	375	17556
5	6	798	349	266	183	23940
8	5	834	289	278	267	33360
2	7	906	445	302	159	12684
7	6	942	373	314	255	39564
4	7	978	457	326	195	27384

In these cases there are half as many solutions as in the previous cases for any given number of distinct prime divisors congruent to 1 mod 3, because there is only a single power of 2 in the perimeters. Notice that we get double solutions with perimeter of 546 and 798, each of which is divisible by two distinct primes congruent to 1 mod 3. Also, for each perimeter with a double solution with p+q even, there is a single solution with half the perimeter with p+q odd.

Anyway, we could specialize still further and consider just the Heronian triangles whose edge lengths are consecutive integers (rather than just in arithmetic progression). For this we need to restrict the value of $|d|$ to be 1. Thus we have

$$p^2 - 3q^2 = \pm 2 \quad \text{if } p+q \text{ is even}$$

$$p^2 - 3q^2 = \pm 1 \quad \text{if } p+q \text{ is odd}$$

These are Pell equations, and can be solved either by the continued fraction for sqrt(3) or by recurrences. The give the infinite families of solutions

Highly Heronian Ellipses

p	q		p	q
1	1		2	1
5	3		7	4
19	11		26	15
71	41		97	56
265	153		362	209
etc.			etc.	

where each sequence of paramaters satisfies the linear recurrence $s[n] = 4s[n-1] - s[n-2]$. These sequences give, alternately, the triangles

p	q	a	b	c	area
1	1	3	4	5	6
2	1	13	14	15	84
5	3	51	52	53	1170
7	4	193	194	195	16296
19	11	723	724	725	226974
26	15	2701	2702	2703	3161340
71	41	10083	10084	10085	44031786

The values of b in this combined sequence satisfy the recurrence $b[n] = 4b[n-1] - b[n-2]$, and the areas satisfy the recurrence $A[n] = 14A[n-1] - A[n-2]$.

12

How Leibniz Might Have Anticipated Euler

In 1671 James Gregory mentioned in correspondence that he had found the inverse tangent of x equals the area under the curve of $1/(1+t^2)$ between $t=0$ and $t=x$. He didn't include his derivation, but it's not too difficult to see how, with the ideas of slopes and quadratures of functions that Gregory and others were developing at the time, this result could be found. Beginning with

$$t = \tan(u) = \frac{\sin(u)}{\cos(u)}$$

we know that the slope of t versus u (in modern notation) is the derivative

$$\frac{dt}{du} = \sin(u)(-1/\cos(u)^2)(-\sin(u)) + (1/\cos(u))\cos(u)$$

$$= \frac{\sin(u)^2 + \cos(u)^2}{\cos(u)^2} = \tan(u)^2 + 1$$

Therefore, since $t = \tan(u)$, we have $dt/du = t^2 + 1$, and so

$$\frac{du}{dt} = \frac{1}{1 + t^2}$$

To invert the function $t = \tan(u)$ we need to express t in terms of u, which we can do by multiplying the above equation by dt and integrating both sides

$$\int du = \int \frac{1}{1 + t^2} dt$$

which is Gregory's result, i.e., we have

$$\arctan(x) = \int_{t=0}^{x} \frac{1}{1 + t^2} dt$$

Indeed, Leibniz derived this same result independently just a few years later (around 1674). Now, by simple division of polynomials he also knew that

$$\frac{1}{1 + t^2} = 1 - t^2 + t^4 - t^6 + t^8 - \ldots$$

Furthermore, although this was prior to development of calculus proper by Newton and Leibniz, several people (including Fermat, Cavalieri, Pascal, etc) had already noticed that the area under the curve t^n from $t=0$ to x is $(x^{(n+1)})/(n+1)$. Thus Gregory had found an infinite series for the inverse tangent of x:

$$\arctan(x) = x - \frac{x^3}{3} + \frac{x^5}{5} - \frac{x^7}{7} + \ldots$$

(Some evidence suggests that this series was actually known in India by about 1500, but the discoverer's identity is not known, nor do we know by what method the result was derived. See Rajagopal, On Medieval Kerala Mathematics, 1986.) A little later, thinking along these same lines, Leibniz noted that since arctan(1)=pi/4 we have the remarkable relation

$$\frac{pi}{4} = 1 - \frac{1}{3} + \frac{1}{5} - \frac{1}{7} + \frac{1}{9} - \frac{1}{11} + \ldots$$

It's interesting to consider what Leibniz (or Gregory) might have done with this result. Unburdened by any concerns about re-arranging the terms of conditionally convergent series, they might well have noticed that the sum could be expressed (at least formally) as a product of geometric series in inverse primes

$$\frac{pi}{4} = \left(1 - \frac{1}{3} + \frac{1}{9} - \ldots\right)\left(1 + \frac{1}{5} + \frac{1}{25} + \ldots\right)\left(1 - \frac{1}{7} + \frac{1}{49} - \ldots\right)\ldots$$

Infinite products were certainly not unknown at this time. For example, Wallis had already given the closely related infinite product

$$\frac{pi}{4} = \frac{2}{3} \cdot \frac{4}{3} \cdot \frac{4}{5} \cdot \frac{6}{5} \cdot \frac{6}{7} \cdot \frac{8}{7} \cdot \frac{8}{9} \cdots$$

Anyway, notice that each geometric series has either alternating signs or is strictly additive, accordingly as p is congruent to -1 or +1 (mod 4). Of course, each of these geometric sums converges, and is given by $1/(1 +- 1/p)$, so throwing cauchy.. er, caution, to the wind, we could express Leibniz's series as the infinite product

$$\frac{pi}{4} = \left(\frac{1}{1 + 1/3}\right)\left(\frac{1}{1 - 1/5}\right)\left(\frac{1}{1 + 1/7}\right)\left(\frac{1}{1 + 1/11}\right)\left(\frac{1}{1 - 1/13}\right)\ldots$$

where the product is taken over all odd primes p, and the sign in the denominator is + or - depending on whether p is congruent

How Leibniz Might Have Anticipated Euler 99

to -1 or +1 modulo 4. This isn't the same as Wallis's product, as can be seen from the first few factors

$$\frac{pi}{4} = \frac{3}{4} \cdot \frac{5}{4} \cdot \frac{7}{8} \cdot \frac{11}{12} \cdot \frac{13}{12} \cdot \frac{17}{16} \cdots$$

Numerically this product appears to converge, albeit very slowly. In any case, considering that Leibniz could digest the notion that $1+2+4+8+.. = -1$ based on the geometric series $1 + x + x^2 + x^3 +..$ equaling $1/(1-x)$, it's hard for me to imagine him being squeamish about this product.

Now here's the interesting part (to me, least). Suppose we ask what happens if we reverse the signs in our infinite product. In other words, what if we use plus signs for primes congruent to -1 (mod 4), and minus signs for primes congruent to +1 (mod 4)? This gives the infinite product

$$\left(\frac{1}{1-1/3}\right)\left(\frac{1}{1+1/5}\right)\left(\frac{1}{1-1/7}\right)\left(\frac{1}{1-1/11}\right)\left(\frac{1}{1+1/13}\right)..$$

which, if we expand it into a sum, is

$$1 + \frac{1}{3} - \frac{1}{5} + \frac{1}{7} + \frac{1}{9} - \frac{1}{11} + \frac{1}{13} - \frac{1}{15} - \frac{1}{17} + \frac{1}{19} + \frac{1}{21} +...$$

where the sign of $1/n$ is plus or minus accordingly as n has an even or odd number of prime divisors (counting multiplicities) congruent to 1 (mod 4). It's easy to form the conjecture based on numerical evidence that this series (and the infinite product) converges on pi/2, i.e., exactly twice the former series (and product).

Now, I don't know of any trigonometric interpretation of this series, analagous to Gregory's arctangent expansion for Leibniz's series, but there might be one. Also, at the risk of introducing some actual math into the discussion, I might just mention Dirichlet's celebrated class number formula

$$\frac{h\,\pi}{2\,\sqrt{|D|}} = \sum_{n=1}^{\inf} \left(\frac{D}{n}\right)\frac{1}{n}$$

where the quantity in parentheses is 0 if n has a common factor with 2D, and otherwise it is the Jacobi symbol. With D=-1 this gives Leibniz's series, showing that the class number h equals 1 for D=-1.

On the other hand, with D=-5 the class number is 2, and Derichlet's formula gives

$$\frac{pi}{\sqrt{5}} = 1 + \frac{1}{3} + \frac{1}{7} + \frac{1}{9} - \frac{1}{11} - \frac{1}{13} - \frac{1}{17} - \frac{1}{19} + \frac{1}{21} + \ldots$$

where the sign of $1/n$ is positive if $n = 1,3,7,9 \pmod{20}$, and negative if $n=11,13,17,19 \pmod{20}$. But I digress.

Returning to our hypothetical Leibniz and his reversal of the signs in the infinite product that formally corresponds to his famous series for pi/4, notice what happens if we multiply his two results together. We immediately have

$$\frac{pi^2}{8} = \left(\frac{1}{1 - 1/3^2}\right)\left(\frac{1}{1 - 1/5^2}\right)\left(\frac{1}{1 - 1/7^2}\right)\left(\frac{1}{1 - 1/11^2}\right)\ldots$$

$$= 1 + \frac{1}{3^2} + \frac{1}{5^2} + \frac{1}{7^2} + \frac{1}{9^2} + \frac{1}{11^2} + \frac{1}{13^2} + \frac{1}{15^2} + \ldots$$

which is the sum of the inverses of all the ODD square numbers.

Furthermore, since it's obvious that the sum of the inverses of the EVEN squares is simply 1/4 times the sum of the inverses of ALL the squares, we have

$$S(all) = S(odd) + S(even) = S(odd) + S(all)/4$$

and so $S(all) = (4/3)\,S(odd)$. (Of course, this also follows simply from multiplying the product involving the odd primes by the factor $1/(1 - 1/2) = 4/3$.) Therefore, our hypothetical (and rather reckless) Leibniz has found that the sum of the inverses of all the squares is $pi^2 / 6$, anticipating Euler and solving the problem that defeated not only the historical Leibniz (judging from his lack of response to Oldenberg's query) but also his disciples the Bernoullis, who after much effort were able to prove only that the sum of the inverse squares converges to a value less than 2, but not to determine the value. It remained for Euler to finally solve the problem, using means that were, if anything, even more reckless and haphazard than the reasoning above (although, to be fair, he did later tidy things up).

13

Odd-Greedy Unit Fraction Expansions

Any ratio of integers m/n can be expressed as a finite sum of unit fractions, i.e., fractions with unit numerators. One method of expanding a given fraction into a sum of unit fractions is via the "greedy algorithm", according to which at each stage we select the largest possible unit fraction (i.e., with the least denominator) less than or equal to the current remainder. It's easy to show that this necessarily terminates in a finite number of steps with a complete unit fraction expansion.

However, if we allow only odd denominators it's not known whether the greedy algorithm applied to a rational number necessarily terminates in a finite number of steps. Of course, we can easily define real numbers that have infinite odd-greedy expansions - just take any infinite sequence of odd unit fractions such that the kthdenominator d[k] is greater than $(1/2)d[k-1]^2 - d[k-1]$ - but we don't know if any such number is rational. (For a more detailed discussion of the possible rationality of infinite odd greedy expansions, see Irrationality of Quadratic Sums.)

Anyway, although most ratios of small numbers can be expressed as a sum of just a few odd unit fractions, it turns out that some of these fractions have surprisingly long odd-greedy expansions. For example, the odd-greedy expansion of

Odd-Greedy Unit Fraction Expansions

3/179 has 19 terms. Moreover, the sequence of numerators of the remainders is striking:

3, 4, 5, 6, 7, 8, 9, 10, 11, 12, 13, 14, 15, 16, 17, 2, 3, 4, 1

This was noted by Stan Wagon, who asked via email if there were fractions with even longer odd-greedy expansions, and for an explanation of the sequence of consecutive remainder numerators, which turns out not to be a unique occurrence. (See The Greedy Algorithm for Unit Fractions.) I suggested to Stan that he look at the fraction 5/139, which also runs to 19 terms, with the sequence of remainder numerators
5, 6, 7, 8, 9, 10, 11, 12, 13, 14, 15, 16, 17, 26, 51, 2, 3, 4, 1

I also directed his attention to a handful of other possibilities that were likely to yield long sequences, with emphasis on 3/2879 and 5/5809 as being strong candidates. Subsequently David Bailey performed an exhaustive computer search for lenghty sequences and re-discovered the fact that these two fractions give unusually long odd-greedy expansions. The fraction 5/5809 is a particularly nice example because the sequence of remainder numerators increases by 1 at each step, all the way to the end: 5,6,7,...,29,30,1.

I also mentioned to Stan (and to David Eppstein) that although the denominators in these sequences can run into the billions of digits, they could be evaluated using modulo arithmetic with a suitable modulus, since the denominators involve only small prime factors. David subsequently applied this approach with the modulus $6(31!)(10!)(6!)(4!)^2$ to compute the remainders of the 5/5809 sequence.

I also told Stan and David that it wouldn't be too difficult to construct fractions with arbitrarily long odd-greedy expansions, but this apparently was overlooked (judging from Richard Guy's subsequent account of the history of this problem in the Dec 98 issue of The American Mathematical Monthly). So I'll show here how to construct a fraction with odd-greedy expansion of length k for any positive integer k.

The key to understanding these expansions is to note that if we begin with any fraction of the form $n/(2m-1)$ where m is divisble by n, the Odd-Greedy algorithm will give the first unit fraction term $1/(2[m/n]+1)$ with the remainder

$$\frac{n+1}{2[(m/n)(2m+n-1)] - 1}$$

so this becomes the new fraction to be expanded. In effect the values of n and m have been transformed according to

$$n \to n+1 \qquad m \to (m/n)(2m+n-1)$$

Now, if our new value of m is divisible by the new value of n, we can repeat the process, and this continues to give the odd greedy terms and remainders until we reach a stage where m is not divisible by n. Notice that the sequence of numerators consists of consecutive integers, and at each stage m is being divided by n and multiplied by the factor (2m+n-1). Therefore, if our initial value of m is something like 100!, it's clear that we can continue this process for numerators up to AT LEAST 100, at which point we will have exhausted the divisibility that we built in to our initial value of m. However, during those 100 steps we have been multiplying m by many large and generally composite numbers, so there's a fair chance that m will have acquired more useful factors by that point.

In any case, we can answer one of our questions by noting that the odd-greedy expansion of 3/(N!-1) must have at least N-2 terms, and the remainders for those terms are 3,4,5,...,N.

Of course, our initial m need not be a factorial, but it must be divisible by the first numerator, and it helps if it has some common factors with some subsequent numerators. To illustrate how this works, consider the case of 3/179, where the initial value of m is 90, which only covers the numerators 3 and 5, with a spare power of 3 and 2. On the very first step we divide m by 3 but multiply by (2m+3-1)=182, which fortuitously gives us another power of 2 to cover the numerator 4 (just in time!), and also gives us 7 and 13, which we will use shortly. Then on the next step we divide by 4 but multiply by 10923, which gives us 11 as well as another factor of 3 (not to mention a factor of 331, which doesn't do us much good directly). As we continue we acquire the factors of 2 and 3 we need to cover the numerators 6, 8, and 9, as well as another factor of 5 to cover 10.

Odd-Greedy Unit Fraction Expansions

This shows how the process tends to feed on itself once it gets started, because the more steps we take, the more factors we apply to m, often providing the factors needed to enable further steps with consecutive remainder numerators. This is the reason long sequences of unit-increasing remainders are so common, even for fairly small fractions.

It's interesting to consider the odd greedy expansion of a fraction such as 3/(500! - 1). We know this will contain at least 500 terms (and since the number of digits in the denominators roughly doubles on each term, the 500th term would have over 2^{500} digits), and the numerators of the remainders will be 4,5,6,...,500, but what will happen then? After cumulatively multiplying into m numbers of the form (2m+n-1) nearly 500 times, we will likely have accumulated a huge number of factors in m, so it wouldn't be too surprising if the sequence continued past the point when the initial m had been exhausted (especially since the terms (2m+n-1) grow exponentially).

Recall how 3/179 essentially "lived off the land" for most of its run. The interesting question is whether there might be some "break-even point", so that a fraction like 3/(500!-1) might have enough initial divisibility to boost it into a sequence where, at least probabilistically, we would expect it to be able to continue generating the factors it needs indefinitely.

I did some checking, and found that for fractions of the form 3/(2m-1) with m=500! we are able to cover 32 of the 64 primes between 500 and 1000. Similarly, with m=100! we are able to cover 10 of the 20 primes between 100 and 200, and with m=50! we can cover 6 of the 10 primes between 50 and 100. I'm not sure off hand why we seem to be able to cover exactly half the primes.

Just for fun, I checked to see how many residues for the initial m value (mod p) are "favorable" for covering p, meaning that by the time we reach a numerator of p we can continue the odd-greedy expansion. Obviously a residue of 0 (mod p) for the initial m guarantees that we can cover p, but I was curious to know how many other residues are favorable for each prime. Here's a little table showing the number of favorable m residues (mod p) for numerators from 1 to 10.

Table: Number of Favorable Initial m Residues (mod p)

Numerator (initial n)

p	1	2	3	4	5	6	7	8	9	10
5	3	3	2	0	0	0	0	0	0	0
7	5	4	4	4	2	0	0	0	0	0
11	5	7	8	9	9	6	4	2	2	0
13	5	5	6	5	4	4	6	6	3	2
17	5	5	7	6	5	3	2	3	4	4
19	1	6	5	4	4	2	7	5	8	8
23	9	6	6	6	6	7	8	9	7	6
29	21	19	21	17	16	16	14	11	8	13
31	27	24	19	18	15	14	15	15	14	12
37	21	18	15	14	13	16	15	18	22	22
41	17	22	20	18	19	19	21	20	22	22
43	17	19	22	24	25	30	33	34	33	33
47	33	33	31	30	33	36	36	32	37	32
53	47	49	47	49	49	45	45	43	47	43
59	31	37	37	35	34	31	27	33	28	17
61	39	33	34	34	28	32	26	29	28	31
67	25	28	29	35	32	26	25	24	32	40
71	19	24	20	24	26	28	28	32	30	22
73	21	18	20	19	17	19	18	28	26	28
79	47	52	47	55	50	47	45	53	53	57
83	41	42	51	58	63	59	59	55	55	53
89	43	29	29	34	39	30	38	46	45	51
97	81	71	71	71	76	71	75	73	77	73
101	21	16	18	24	24	30	23	23	32	35
103	21	22	21	22	16	16	12	9	10	12
107	63	58	50	46	42	44	45	55	54	49
109	7	4	2	4	6	4	4	4	9	12
113	81	85	85	87	85	91	97	93	88	89
127	93	100	91	95	91	91	97	91	91	88
131	31	33	33	34	29	33	34	36	40	40
137	63	64	68	64	66	69	71	71	66	66
139	73	79	85	90	94	100	94	91	93	103
149	105	103	103	96	93	95	92	92	89	87

Odd-Greedy Unit Fraction Expansions

151	17	20	19	17	14	14	10	10	8	15
157	29	24	25	28	34	36	46	47	46	48
163	47	40	40	39	35	34	30	30	24	24
167	25	22	20	23	20	18	17	16	14	14
173	97	95	92	97	94	88	76	77	75	80
179	151	153	155	155	155	151	149	148	145	144
181	173	175	175	177	177	177	179	179	179	175
191	109	104	96	84	85	77	81	82	84	82
193	47	44	40	48	45	59	59	55	54	44
197	93	100	89	88	86	78	79	83	80	87
199	161	163	159	148	143	148	148	141	133	128

The average ratio of favorable residues to total number of residues seems to be close to 1/2. Also, notice that 109 is evidently a "sticky wicket", especially for fractions with numerator 3, because the only way past is to either make the initial m value a multiple of 109 or make it congruent to -1 (mod 109). In contrast p=181 has over 96% favorable.

Factoring Zeta

We've seen how Leibniz might have anticipated Euler by discovering that the sum of the inverse squares is (pi^2)/6 purely on the basis of the arctangent series for pi/4. Another way of expressing that approach is to regard it as essentially a factorization of the product form of zeta(2). In other words, beginning with the expression.

$$1 + \frac{1}{2^2} + \frac{1}{3^2} + \frac{1}{4^2} + \frac{1}{5^2} + \ldots = \text{PROD}_{\text{all primes}} \left(\frac{1}{1 - 1/p^2} \right)$$

we factor the right hand side into two parts

$$\text{PROD}\left(\frac{1}{1 - 1/p^2} \right) = \text{PROD}\left(\frac{1}{1 - 1/p} \right) \text{PROD}\left(\frac{1}{1 + 1/p} \right)$$

However, the two individual products on the right side aren't very useful because the first is just the harmonic series, which diverges to infinity, and the second converges on zero. In order to get a more useful factorization, we note that the above is really just a special case of a more general form

$$\text{PROD}\left(\frac{1}{1 - 1/p^2}\right) = \text{PROD}\left(\frac{1}{1 - a(p)/p}\right) \text{PROD}\left(\frac{1}{1 - b(p)/p}\right)$$

where we've replaced the unit numerators with the functions $a(p)$ and $b(p)$, which for each prime p are the roots of $x^2 - 1$, but any permutation of those roots is allowable for any given prime. For example, we are free to define $a(p)$ and $b(p)$ in terms of the Legendre symbol

$$a(p) = -b(p) = \left(\frac{-1}{p}\right) = \begin{pmatrix} +1 & \text{if } p=1,2 \bmod 4 \\ -1 & \text{if } p=3 \bmod 4 \end{pmatrix}$$

in which case we have decomposed $\text{zeta}(2)$ into the factors

$$\text{PROD}\left(\frac{1}{1 - a(p)/p}\right) \text{PROD}\left(\frac{1}{1 - b(p)/p}\right) = \frac{\pi}{2} \frac{\pi}{3}$$

where the first factor is simply $1/(1 - 1/2)$ times the arctan series for $\pi/4$, and the second factor is $1/(1 + 1/2)$ times the sign-reversal of the Leibniz series (product) as discussed in the previous note.

Now, what happens if we take this same approach to $\text{zeta}(3)$, i.e., the sum of the inverse CUBES? There is no known "simple" expression for zeta with odd arguments, in contrast with the fact that all the even-index zeta values are given by

$$\text{zeta}(2n) = \frac{(2\pi)^{(2n)}}{2(2n)!} |B_{\{2n\}}|$$

where $B_{\{2n\}}$ is the $(2n)$th Bernoulli number. The problem with this formula relative to odd zeta arguments is that all the odd-index Bernoulli numbers vanish. (Of course, the odd-index

Odd-Greedy Unit Fraction Expansions

Bernoulli POLYNOMIALS do not vanish, so people have tried to generalize the above expression in terms of the full polynomials rather than just the constant coefficients, but nothing seems to work.)

Anyway, following our factorization of zeta(2), we might try the same trick on the product-form of zeta(3), i.e., we might try to factor the product

$$\text{zeta}(3) = \text{PROD}\left(\frac{1}{1 - 1/p^3} \right)$$

which can obviously be formally factored as

$$\text{PROD}\left(\frac{1}{1 - 1/p^3} \right) = \text{PROD}\left(\frac{1}{1 - 1/p} \right) \text{PROD}\left(\frac{1}{1 + 1/p + 1/p^2} \right)$$

but again we have the problem that the first factor is the harmonic series (divergent) and the second factor converges to zero. We might simply replace the "1/p" terms with a(p)/p and b(p)/p, but it seems more in the spirit of our previous example to factor zeta(3) into LINEAR factors.

To simplify the notation, let me use square brackets to denote a product evaluated over all primes, so our putative factorization of zeta(3) can be written a

$$\left[\frac{1}{1 - 1/p^3} \right] = \left[\frac{1}{1 - a(p)/p} \right]\left[\frac{1}{1 - b(p)/p} \right]\left[\frac{1}{1 - c(p)/p} \right]$$

where a(p), b(p), and c(p) are now (some permutation of) the roots of $x^3 - 1$. Except for the prime 3, every prime falls in one of the congruence classes {1,2}, {4,5}, {7,8} modulo 9, so if we let r1,r2,r3 denote the cube roots of 1

$$r1 = 1 \qquad r2 = (-1+\sqrt{-3})/2 \qquad r2 = (-1-\sqrt{-3})/2$$

then one possible way of defining the functions a,b,c is presented below, along with the resulting factorization of zeta(3).

	p (mod 9)			
	1,2	4,5	7,8,3	product
a(p)	r1	r2	r3	F1 = 1.484093 - 0.209977 i
b(p)	r2	r3	r1	F2 = 1.057036 + 0.224225 i
c(p)	r3	r1	r2	F3 = 0.740445 - 0.050781 i

Multiplying the factors together gives

$$\text{zeta}(3) = F1 * F2 * F3 = 1.20205...$$

Unfortunately, unlike the case of zeta(2), these "linear factors" aren't immediately recognizable as fractions of pi (or anything else). However, there were quit a few arbitrary choices made in the above construction, mainly involving the definitions of the functions a,b,c. In the case of zeta(2) we essentially classified the primes into two categories according to whether $x^2 + 1$ splits or is irreducible in the field of integers modulo p (i.e., whether or not -1 is a quadratic residue), so it would seem that for zeta(3) we should classify the primes into three categories according to whether some cubic polynomial splits completely, has one linear factor, or is irredicible in Z_p. However, we can't use $x^3 + 1$ because that ALWAYS has at least one linear factor. It's not clear to me how to come up with the "most natural" factorization of zeta(3).

By the way, it's possible that we should really work with 3! = 6 factors, maybe using +- the cube roots of 1. This may be related to the fact that we can get zeta(3) from EITHER of the products $1/(1 - 1/p^3)$ or $1/(1 + 1/p^3)$, because they are "conjugate" factors of zeta(6), i.e.,

$$\left| \frac{1}{1 - 1/p^6} \right| = \left| \frac{1}{1 - 1/p^3} \right| \left| \frac{1}{1 + 1/p^3} \right| = \frac{pi^6}{945}$$

So if we had a simple expression for either of the factors we would automatically have one for the other factor. Unfortunately the second factor is just as intractible as the first, so this doesn't seem to help much, except perhaps to suggest that the cube roots of -1 should also be included in a six-term factorization.

14

Four Squares from Three Numbers

Diophantus posed the problem of finding three rational numbers a,b,c such that ab+1, ac+1 and bc+1 are each squares, and this problem has been extended and generalized (and specialized) in several different ways by later mathematicians. For example, Euler found sets of four, and one set of FIVE, rationals numbers such that each of the pairwise products is one less than a square. A discussion of this general problem can be found in If ab+1, ac+1, bc+1 are squares.... There's also an article in the Feb 98 Mathematics Magazine on this subject.

Anyway, John Gowland recently proposed a slightly different extension of this problem, namely, to find three integers such that all FOUR of the quantities ab+1, ac+1, bc+1, abc+1 are distinct squares, i.e., we require integer solutions of the simultaneous equations

$$ab+1 = C^2$$
$$ac+1 = B^2$$
$$bc+1 = A^2$$
$$abc+1 = D^2$$

Obviously any such triple is a solution to the original problem of Diophantus, but the 4th condition involving the product of all three numbers significantly restricts the set of solutions.

Notice that we stipulated "distinct" squares, because some of the infinitely many solutions of Diophantus's original problem have a=1, and since bc+1 is a square they automatically satisfy abc+1=square as well. To avoid these trivial solutions we require all four of the squares to be distinct. (Equivalently, each of the numbers a,b,c must be greater than 1.)

To characterize the solutions of this extended problem, it might be useful to recall Saunderson's parameterization giving a large class of solutions a,b,c of the original 3-square problem:

$$a = n \qquad b = q(qn+2) \qquad c = (q+1)[(q+1)n+2]$$

where n is an integer and q is any RATIONAL number such that b and c are integers. For example, with n=45 and q=2/5 we have

$$a=45 \qquad b=8 \qquad c=91$$

This example wan't chosen entirely at random, because it happens to also be a solution of the extended problem, as shown by the fact that

$$(45)(8)(91) + 1 = 181^2$$

However, not surprisingly, only a small fraction of the solutions given by Saunderson's parameterization are also solutions of the extended problem. Notice that if we substitute Saunderson's expressions for a,b,c into the equation abc+1 = D^2 we have a quartic in q which happens to be fairly easy to solve explicitly, giving the result

$$q = \frac{-(n+2) +- \sqrt{n^2 + 4 +- 4\sqrt{n(n + D^2 - 1)}}}{2n}$$

Now, we need q to be rational, so the quantity under the radical must be a square. Thus we have an integer w such that

$$w^2 = n^2 + 4 +- \sqrt{n(n + D^2 - 1)}$$

and this requires that the quantity in the square root must be a square, so we have an integer m such that

Four Squares from Three Numbers

$$m^2 = n^2 + n(D^2 - 1)$$

Solving this for n leads us to conclude that $(D^2-1)^2$ plus $4m^2$ must be a square, so we have a Pythagorean triple

$$(D^2 - 1)^2 + (2m)^2 = R^2$$

for some integer R. Obviously this implies the existence of an integer k and mutually coprime integers x,y (one odd, one even) such that

$$D^2 - 1 = 2kxy \qquad 2m = k(x^2 - y^2) \qquad R = k(x^2 + y^2)$$

or possibly swapping the first two expressions depending on whether $2m/k$ is odd or even. Here's a little table summarizing the smallest solutions of the full 4-square problem

a	b	c	D	k	x	y
5	7	24	29	10	7	6
45	8	91	181	90	14	13
84	20	186	559	168	31	30
102	44	280	1121	204	56	55
105	8	171	379	210	19	18
119	40	297	1189	238	55	54
133	3	176	265	266	12	11
105	11	184	461	210	23	22
301	24	495	1891	602	55	54

Several regularities are apparent in this table, e.g., the value of k is always equal to 2a, and we can verify that m always equals $a(x+y)$, which is obvious because $x-y = 1$ in all cases.

If we take advantage of these regularities, the problem reduces to finding integers n and x such that

$$D^2 = 4nx(x-1) + 1$$
$$w^2 = n^2 + 4 +- 4n(2x-1)$$
(1)

Given such integers, the value of Saunderson's q is then

$$q = -\frac{n+2 +- w}{2n}$$

from which we can compute the values of the triple a,b,c. Here is a table of all the solutions given by this form for n and x less than 10000:

n	x	a	b	c	A	B	C	D
5	7	5	7	24	13	11	6	29
45	14	8	45	91	64	27	19	181
3	77	3	133	176	153	23	20	265
105	19	8	105	171	134	37	29	379
11	70	11	105	184	139	45	34	461
20	63	20	84	186	125	61	41	559
44	85	44	102	280	169	111	67	1121
119	55	40	119	297	188	109	69	1189
301	55	24	301	495	386	109	85	1891
477	66	24	477	715	584	131	107	2861
85	456	85	672	1235	911	324	239	8399
114	561	114	816	1540	1121	419	305	11969
165	498	165	664	1491	995	496	331	12781
132	575	132	820	1610	1149	461	329	13201
280	469	280	546	1608	937	671	391	15679
2387	175	40	2387	3045	2696	349	309	17051
74	2065	74	3612	4720	4129	591	517	35519
85	6432	85	11859	13952	12863	1089	1004	118591
120	9880	120	18278	21360	19759	1601	1481	216449
3193	4182	3193	4551	15368	8363	7005	3812	472565
5705	7258	5705	7831	26904	14515	12389	6684	1096339

An exhaustive search of all integer triples a,b,c satisfying all four conditions doesn't seem to turn up any additional solutions (in this range) beyond those listed here, so it's possible that (1) represents the necessary and sufficient conditions.

Of course, there are many symmetries here, beyond what I've explicitly used, and I've just listed the first (n,x) yielding each solution triple, but there are always at least three such pairs that give the same triple (corresponding to the fact that we can identify the parameter n with a, b, or c).

Four Squares from Three Numbers

One striking feature of the table is how, when arranged by increasing D, the values of n occur in "sawtooth" cycles, with the following values

```
 5   45
 3  105
11   20   44  119  301  477
85  114  165  132  280 2387
74   85  120 3193 5705 ...
```

I suppose this might just be an artifact of that way I've selected representative solutions and arranged them, but it seems to suggest some underlying pattern that hasn't been brought out explicitly.

By the way, this problem also relates to the subject of representing numbers as products of "shy squares" in multiple ways. (A "shy square is one less than a square). Solutions to the general problem above obviously represent integers that can be factored in two distinct ways into shy squares, namely

$$(A^2 - 1)(B^2 - 1)(C^2 - 1) = (D^2 - 1)^2$$

In other words, this is a product of three distinct shy squares that is also the square of a shy square. There are several open questions in this area. For example, it's possible for a number to be the product of two shy squares in 5 distinct ways, but it's not known if there exists a six-way expressible number.

Meandering Convergence of a Dirichlet Series

In a previous note we discussed how Leibniz might have anticipated Euler's summation of the inverse square integers by factoring the product form of the arctan(1) series. This leads to the interesting "Dirichlet" series

$$\frac{pi}{2} = 1 + \frac{1}{3} - \frac{1}{5} + \frac{1}{7} - \frac{1}{9} + \frac{1}{11} - \frac{1}{13} - \frac{1}{15} + \frac{1}{17} - \frac{1}{19} + \frac{1}{21} - \cdots$$

$$= \sum_{j=0}^{\inf} \frac{(-1)^{f(2j+1)}}{2j+1}$$

where f(k) is the sum of the exponents of primes congruent to 1 modulo 4 in the prime factorization of k. If we let s(x) denote the partial sum of this series up to 2j+1 < x we find that it fairly quickly achieves over 99% of pi/2, but then it meanders around for a very long time, occassionally coming very close to pi/2 (from below) and then backing off. (By comparison, the series for arctan(1) converges FAR more rapidly and uniformly on pi/4.)

Just for fun, I plotted the difference pi/2 - s(x) versus x for values of x up to 10 million, as shown below:

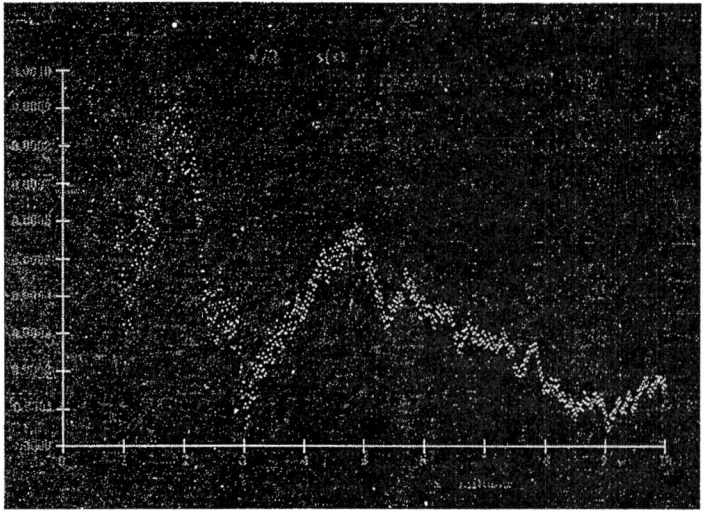

Interestingly, the difference between s(x) and pi/2 dips at about 1/3 million, then again at about 1 million, again at 3 million, and then again at very close to 9 million. I haven't checked, but it seems likely that the next near-approach to zero would occur around 27 million, and so on. This seems to suggest that the parity of the exponents of prime factors congruent to 1 (mod 4) exhibits a distinct logarithmic periodicity in the sequence of natural numbers.

Is this periodicity genuine? If so, what causes it, and what is the precise frequency? Also, is it true that pi/2 - s(x) is always positive? If not, at what value of x does this difference first go negative?

15

Accidental Melodies

The range of acoustic tones perceivable by humans extends roughly from 20 Hz up to 20,000 Hz (cycles per second), which is a factor of 1000. Thus, noting that $2^{10} = 1024$, the audible range of frequencies represents about 10 doublings. The harmonic association of tones with doubled or halved frequencies is so strong that we tend to regard all tones with frequencies $2^n v$ as "the same" in some sense. As a result, we conceive of the audible range of tones in terms of 10 "copies" of the same sequence of tones, each of which covers frequencies up to a factor of 2, say from v up to 2v for some arbitrary fixed frequency v.

Within each of these ranges there exist other harmonic associations, not as strong as the 2:1 harmonic, but strong enough to make the combinations of certain tones aesthetically pleasing. The most notable is the 3:2 harmonic. (Incidentally, this harmonic features in the gravitational coupling between the rotations and revolutions of the planet Mercury.) Beginning with some basic frequency v, we can increase this successively by the ratio 3/2, and divide by factors of 2 as necessary to express the results in the range 1 to 2. This gives the sequence of frequencies in the left-hand columns below. On the other hand, beginning from the same basic frequency, we can successively divide by 3/2, and multiply by factors of 2 as necessary to express the results in the range 1 to 2. This gives the sequence of frequencies listed in the right hand columns.

Ascending 3/2 Progression		
1 /	1	1.000000
3 /	2	1.500000
9 /	8	1.125000
27 /	16	1.687500
81 /	64	1.265625
243 /	128	1.898438
729 /	512	1.423828
2187 /	2048	1.067871
6561 /	4096	1.601807
19683 /	16384	1.201355
59049 /	32768	1.802032
177147 /	131072	1.351524
531441 /	524288	2.027286

Descending 3/2 Progression		
1 /	1	1.000000
4 /	3	1.333333
16 /	9	1.777777
32 /	27	1.185185
128 /	81	1.580247
256 /	243	1.053498
1024 /	729	1.404664
4096 /	2187	1.872885
8192 /	6561	1.248590
32768 /	19683	1.664787
65536 /	59049	1.109858
262144 /	177147	1.479811
1048576 /	531441	1.973081

In each of these sequences, on the 12th step we arrive nearly at twice the initial base tone, differing by only about 1.36%. (I've allowed the last entry in the left hand columns to exceed 2, since it is very nearly equal to 2.) This occurs because 3^{12} and 2^{19} happen to be nearly equal. If we re-arrange the entries in these columns so that the frequencies are in ascending order, we get the results shown below.

Ordered 3/2 Ascending Progression		
1 /	1	1.000000
2187 /	2048	1.067871
9 /	8	1.125000
19683 /	16384	1.201355
81 /	64	1.265625
177147 /	131072	1.351524
729 /	512	1.423828
3 /	2	1.500000
6561 /	4096	1.601807
27 /	16	1.687500
59049 /	32768	1.802032
243 /	128	1.898438
531441 /	524288	2.027286

Ordered 3/2 Descending Progression		
1 /	1	1.000000
256 /	243	1.053498
65536 /	59049	1.109858
32 /	27	1.185185
8192 /	6561	1.248590
4 /	3	1.333333
1024 /	729	1.404664
262144 /	177147	1.479811
128 /	81	1.580247
32768 /	19683	1.664787
16 /	9	1.777777
4096 /	2187	1.872885
1048576 /	531441	1.973081

These two harmonic progressions give two very similar

Accidental Melodies

sequences of 12 tones. If we take the geometric mean of each pair of corresponding tones, we get the sequence of squared frequencies listed below, along with the ratios of consecutive frequencies.

Squared mean frequencies	frequency ratios
$2^0 / 3^0$	
$3^2 / 2^3$	$3/\sqrt{8}$
$2^{13} / 3^8$	$2^8 / 3^5$
$3^6 / 2^9$	$3^7 / 2^{11}$
$2^7 / 3^4$	$2^8 / 3^5$
$3^{10} / 2^{15}$	$3^7 / 2^{11}$
$2^1 / 3^0$	$2^8 / 3^5$
$2^{17} / 3^{10}$	$2^8 / 3^5$
$3^4 / 2^5$	$3^7 / 2^{11}$
$2^{11} / 3^6$	$2^8 / 3^5$
$3^8 / 2^{11}$	$3^7 / 2^{11}$
$2^5 / 3^2$	$2^8 / 3^5$
$2^2 / 3^0$	$3/\sqrt{8}$

There are only three distinct ratios between consecutive tones in this sequence, namely

$$\frac{3}{2\sqrt{2}} = 1.060660 \qquad \frac{2^8}{3^5} = \frac{256}{243} = 1.053498 \qquad \frac{3^7}{2^{11}} = \frac{2187}{2048} = 1.067871$$

To the approximation that $2^{19}/3^{12} \sim 1$, each of these represents an approximation of the 12th root of 2, as shown by the identities

$$\left(\frac{3}{2\sqrt{2}}\right)^{12} = 2\left(\frac{3^{12}}{2^{19}}\right) \qquad \left(\frac{2^8}{3^5}\right)^{12} = 2\left(\frac{2^{19}}{3^{12}}\right)^5 \qquad \left(\frac{3^7}{2^{11}}\right)^{12} = 2\left(\frac{3^{12}}{2^{19}}\right)^7$$

Therefore it's not surprising that the (normalized) ascending and descending powers of 3/2 closely bracket the geometric progression of the powers of $2^{1/12}$, as shown in the table below.

n	Descending 3/2 Powers	$[2^{1/12}]^n$	Ascending 3/2 Powers
0	1.000000	1.000000	1.000000
1	1.053498	1.059463	1.067871
2	1.109858	1.122462	1.125000
3	1.185185	1.189207	1.201355
4	1.248590	1.259921	1.265625
5	1.333333	1.334840	1.351524
6	1.404664	1.414213	1.423828
7	1.479811	1.498307	1.500000
8	1.580247	1.587401	1.601807
9	1.664787	1.681793	1.687500
10	1.777777	1.781797	1.802032
11	1.872885	1.887748	1.898438
12	1.973081	2.000000	2.027286

In addition to the 3/2 harmonic, several other fundamental harmonics are approximated by elements of these sequences. The most prominent are the tones with n = 0, 2, 4, 5, 7, 9, 11, and 12, which correspond approximately to the harmonic ratios 1, 9/8, 5/4, 4/3, 3/2, 5/3, 15/8, and 2 respectively. These comprise the basic 8-tone scale, with frequencies proportional to

1	9/8	5/4	4/3	3/2	5/3	15/8	2
1.000	1.125	1.250	1.333	1.500	1.667	1.875	2.000
(0)	(2)	(4)	(5)	(7)	(9)	(11)	(12)

The numbers in parentheses signify the tone from the previous table that most nearly approximates each of these frequencies. Clearing the fractions, the frequencies are proportional to

$$24 \quad 27 \quad 30 \quad 32 \quad 36 \quad 40 \quad 45 \quad 48$$

The figure below shows the logarithms of the (normalized) ascending and descending progression of powers of 3/2, and the powers of the 12th roots of 2, and the eight harmonic tones.

Accidental Melodies

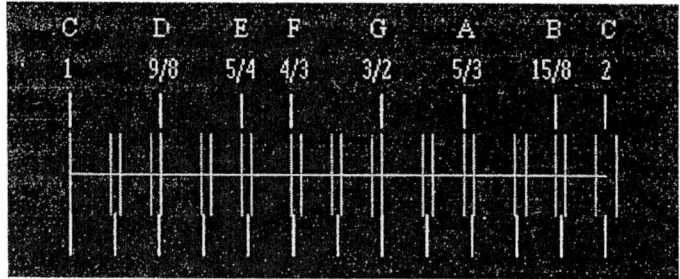

Since this is a logarithmic scale, the 12th roots of 2 are uniformly spaced on the interval. Using these tones rather than the exact harmonic rational fractions gives the "equally-tempered" scale used in most modern music, particularly keyboard music. This strictly geometric scale was advocated by Bach and others in the 18th century to enable smooth modulation from one key to the next in the course of a single composition. It might have been more descriptively named the logarithmic scale.

Since the 12th root of 2 is irrational, none of these terms in the "equally-tempered" progression correspond exactly with any rational harmonics. However, each number of the form $2^{k/12}$ for $k = 0, 1, ..., 12$ can be expressed as a continued fraction of the form

$$2^{k/12} = a_0 + \cfrac{1}{a_1 + \cfrac{1}{a_2 + \cfrac{1}{a_3 + \cfrac{1}{...}}}}$$

The first several coefficients for the continued fractions of the twelve equally-tempered semi-tones are listed below.

```
 0:  1   0   0   0   0   0   0   0   0   0   0   0   0   0   0   0   0   0   0
 1:  1  16   1   4   2   7   1   1   2   2   7   4   1   2   1  59   1   3   2
 2:  1   8   6  31   1   2   2   2  10   3   1   9   1   2  16   1   1 310   1
 3:  1   5   3   1   1  40   5   1   1  25   2   3   1   6   2   2   1   1   4
 4:  1   3   1   5   1   1   4   1   1   8   1  14   1  10   2   1   4  11   1
 5:  1   2   1  73  11   1   1  10   2   9 318   1   1   1   4   1  11   3   6
 6:  1   2   2   2   2   2   2   2   2   2   2   2   2   2   2   2   2   2   2
 7:  1   2 147   5   1   3   5   4   4   1   1 159   1   1   2   6  13   2   1
 8:  1   1   1   2   2   1   3   2   3   1   3   1  30   1   4   1   2   9   5
 9:  1   1   2   7  81   2   1   3  12   1   2   1   1   2   3   5   1   4   3
10:  1   1   3   1   1   2   1   1  15   2   1   4   1  20   1   1  20   2   1
11:  1   1   7   1   9   1  15   5   1  14   2   2   1   2  29   1   1   1  27
12:  2   0   0   0   0   0   0   0   0   0   0   0   0   0   0   0   0   0   0
```

The eight harmonic tones are given by the first or second convergents, i.e., for n = 0, 2, 3, 5, 7, 9, 11, and 12 we have [1;0], [1;8], [1;3,1], [1;2,1], [1;2], [1;1,2], [1;1,7], [2;0].

Our familiar eight-tone major scale is really just one of "modes" that were developed in Western Europe beginning in the Middle Ages. They were originally defined by the Roman Catholic Church as part of the singing of the psalms. Each mode had an eight-tone scale with a designated "reciting note", around which most of the words were sung (or chanted) and a "final note" on which the song was to end. The full set of seven modes (not counting the "hypo", aka plagal, modes, which differed only in range) were as listed below

Aeolian	**A** B C D $\underline{E}$ F G A
Locrian	**B** C D E $\underline{F}$ G A B
Ionian	**C** D E F $\underline{G}$ A B C
Dorian	**D** E F G $\underline{A}$ B C D
Phrygian	**E** F G A $\underline{B}$ C D E
Lydian	**F** G A B $\underline{C}$ D E F
Mixolydian	**G** A B C $\underline{D}$ E F G

For each mode the "finalis" is listed in bold type, and the reciting note (also called the tenor) is underlined. The original formulation of musical modes for church music is attributed to St. Ambrose (c. 340-397), who set down the rules for what came to be called the Dorian, Phyrgian, Lydian, and Mixolydian modes. Interestingly, the Ionian and Aoelian modes were apparently not formalized until the 16th century, by a Swiss monk named Henry of Glarus (1488-1563). (The Locrian mode was hardly ever used, because of its unpleasant tonic chord, and is included on the

Accidental Melodies

list only for completeness.) Of all these seven modes, only two survive today, the Ionian and Aeolian, which we call the major and minor scales.

Most common melodies make use of only tones that are among the seven main tones (up to powers of 2) of the key in which they are written. For example, the well-known tune "Row, Row, Row Your Boat" is written in the major mode, whose main tones from the full 12-tone scale are 0, 2, 4, 5, 7, 9, 11, and the melody consists of the sequence of tones

```
 0  0  0  2  4
 4  2  4  5  7
12  7  4  0
 7  5  4  2  0
```

The note n = 12 is just double the frequency if n = 0, so we don't count this as a distinct tone. Hence this melody uses just five of the seven distinct tones in the major key, namely, 0, 2, 4, 5, and 7. More sophisticated melodies make use of all seven of the major tones in the chosen key.

Furthermore, a composer sometimes calls for a note that is not even in the chosen key. Such a note is prefixed with its own sharp (#) or flat ($) symbol, and it's called an *accidental*. One example of a "highly accidental" melody is the title song from the 1963 movie "From Russia With Love" (composed by Lionel Bart), which consists of the following sequence of tones:

```
 0  5  8  5 13 12 11 12  7
 0  5  8  5 15 14 13 12 15
12 17 12  8  5 15 15 13 10
 7 13 12 16 13 12 17
```

Examining this modulo 12, we find that there is at least one appearance of each of the tones 0, 1, 2, 3, 4, 5, 7, 8, 10, and 11. In other words, of the twelve distinct semi-tones in the full equally-tempered scale, this melody invokes ten of them (the missing tones are 6 and 9), so it is what I would call a highly accidental melody.

An even more impressive example is "The Shadow of Your Smile", composed by Johnny Mandel for the 1965 movie "The Sandpipers". This song won the Academy Award for Best Original

Song, presumably because the Academy members were appreciative of the fact that the melody invokes eleven of the twelve tones in the full equally-tempered scale. The sequence of tones in this melody is

```
 0  5  7  8 12  2  7  4  1  0
 0  5  7  8 12  5 10  7  3 12
13 12 10  8 10  7  0  7 10 10  9  8  5  0
 8 12 10  8  7  8  0 -1  8  7
```

Again evaluating these tones modulo 12, we see that the only one of the twelve tones in the scale not touched is $n = 6$. Naturally the works of Bach include compositions in which he systematically progressed through each semi-tone of the equally-tempered scale, but I'm not aware of any simple popular melody that contains all twelve semi-tones.

16

Some Properties of the Lucas Sequence

The second-order recurrence

$$s[n] = 4\,s[n-1] - s[n-2] \qquad (1)$$

arises in many different situations, such as in the sequence of "Lucas Numbers" used in primality tests of Mersenne numbers, and in Heronian triangles, and continued fractions for sqrt(3), and in general anything that involves the Pell equation $x^2 - 3y^2 = 4$.

With the "natural" initial values $s[0]=2$ and $s[1]=4$, we have the sequence

2   4   14   52   194   724   2702   10084   37634   ...

This sequence comprises the integer values of s that satisfy the Pellian equation

$$s^2 - 3r^2 = 4 \qquad (2)$$

with the corresponding values of r listed below

s	r
2	0

4	2
14	8
52	30
194	112
724	418
etc.	

I referred to (2) as "Pellian" rather than a Pell equation proper, because the right hand side is not unity. However, given any solution of the Pellian equation we can obviously construct others by simple multiuplication of similar quadratic forms. For example, suppose we have integers s,r that represent a solution of $s^2 - 3r^2 = 4$. Then for any of the infinitely many integer pairs u,v that satisfy the Pell equation $u^2 - 3v^2 = 1$ we have

$$4 = (u^2 - 3v^2)(s^2 - 3r^2) = (us +- 3vr)^2 - 3(ur +- vs)^2$$

Now, the minimal non-trivial solution of $u^2 - 3v^2 = 1$ is (2,1), so beginning with s[0]=2, r[0]=0 we can generate successive solutions of $s^2 - 3r^2 = 4$ by the recurrence

$$s[n] = 2\ s[n-1] + 3\ r[n-1]$$

$$r[n] = 1\ s[n-1] + 2\ r[n-1]$$

Naturally the characteristic polynomial for the right hand matrix is simply the characteristic polynomial

$$f(x) = x^2 - 4x + 1$$

of the 2nd-order recurrence (1), and that recurrence itself can be recovered by eliminating r from the two first-order equations.

Torsten Sillke noticed, based on numerical evidence, that if p > 3 is a prime divisor of s[n] - 1 then apparently p is necessarily congruent to 1 or 17 (mod 24) if n is odd, and it is congruent to 1 or 13 (mod 24) if n is even. He asked if this is true in general and, if so, for a proof.

To prove that this conjecture is correct, let's re-write the basic Pellian equation in the form

Some Properties of the Lucas Sequence

$$s^2 - 4 = 3r^2$$

which can be factored to give

$$(s+2)(s-2) = 3r^2$$

By construction, s[n] is divisible by exactly one power of 2 if n is even, and by two powers of 2 if n is odd. Now, if n is even, there is an odd integer y such that y=s/2, and we have

$$(y+1)(y-1) = 3r^2$$

where the two factors on the left are coprime except for a single common factor of 2. Also, from congruence considerations we know that $s/2 + 1$ is not divisible by 3, so we have integers a,b such that

$$(y+1)/2 = a^2 \qquad (y-1)/2 = 3b^2$$

which implies that

$$s[n]-1 = (2a)^2 - 3 = 1 + 3(2b)^2 \qquad \text{(n even)}$$

Notice that these are special cases of the quadratic forms

$$x^2 + Ny^2$$

where gcd(x,Ny)=1. Euler (and probably Fermat) proved that, for the cases N=1, +-2, 3, every prime divisor of this form is also expressible in this form, and the prime p can divide a number of this form if and only if (-N) is a square modulo p. Thus, any prime divisor of a number of the form $a^2 - 3$ (with a prime to 3) must be congruent to 1, 11, 13, or 23 modulo 24, because those are the only primes modulo which 3 is a square. Likewise, any prime divisor of a number of the form $1 + 3b^2$ must be congruent to 1, 7, 13, or 19 modulo 24, because those are the prime modulo which -3 is a square. Now, since s(n)-1 must be of BOTH those forms if n is even, it's clear that any prime divisor must be in the intersection of those two sets of residues, so it can only be 1 or 13 (mod 24).

For the case of odd n there is an odd integer y such that $2y = s/2$, and we have

$$(2y+1)(2y-1) = 3(2r)^2$$

The two factors on the left are coprime, so one must be a square and the other must be 3 times a square. It's easy to show that $s[n]/2 - 1$ is not divisible by 3 if n is odd, so we have integers a,b such that

$$2y + 1 = 3a^2 \qquad 2y - 1 = b^2$$

from which it follows that

$$s[n]-1 = 3(2a^2 - 1) = 2b^2 + 1 \qquad (n \text{ odd})$$

So in this case, after factoring out the prime 3, we need only consider the prime divisors of numbers that have BOTH the forms

$$x^2 - 2y^2 \qquad \text{and} \qquad x^2 + 2y^2$$

with x coprime to 2y. It follows that any prime divisor of $2a^2 - 1$ must be a prime p modulo which 2 is a square, so p must be congruent to 1, 7, 17, or 23 (mod 24). Also, any prime divisor of $2b^2 + 1$ (other than 3) must be a prime p modulo which -2 is a square, so p must be congruent to 1, 3, 11, 17, or 19 (mod 24). Again, since s(n)-1 must be of BOTH those forms if n is odd, it's clear that any prime divisor must be in the intersection of those two two sets of residues, so it can only be 1 or 17 (mod 24) (once we have factored out the power of 3). This completes the proof of Sillke's conjecture.

Incidentally, without even invoking the theory of quadratic forms we can nearly prove the above result by simple congruence considerations. We know that s[n]-1 is congruent to 3 (mod 24) for all odd indicies n, so it's obvious that the product of the remaining factors of (s[n]-1)/3 must be either 1 or 17. (We can rule out 9 because $2a^2 - 1$ cannot be congruent to 9 modulo 24). Therefore, s[n]-1 = 3m where m = 1 or 17. If m is a prime, then we are done. If m is composite, the question is whether

Some Properties of the Lucas Sequence

each of the prime factors of m is necessarily congruent to 1 or 17 (mod 24).

If we examine the multiplication table (mod 24) we find that if m = uv then the only possible values of the factors u and v are (mod 24)

$$1 = 1*1 = 5*5 = 7*7 = 11*11 = 13*13 = 17*17 = 19*19 = 23*23$$
and
$$17 = 1*17 = 5*13 = 7*23 = 11*19$$

Therefore, any prime factors of m must be in one of these residue classes. However, we can exclude many of these, recalling the fact that

$$V(n+2) = 7V(n) + 4 \text{ sqrt}[V(n)^2 - 12]$$

This shows that if there exists an index j such that $V(j)$ is congruent to 1 (mod p) then

$$V(j+2) = 7 + 12 \text{ sqrt}(-1) \quad (\text{mod } p)$$

which implies that -1 must be a square modulo p, which means that p = 1 (mod 4). Thus we can rule out p = 7, 11, 19, or 23 (mod 24), which just leaves the following possibilities for any 2-part factorizations of m (mod 24)

$$1 = 1*1 = 5*5 = 13*13 = 17*17$$
and
$$17 = 1*17 = 5*13$$

So, the only part of Sillke's conjecture (for odd n) that requires the theory of quadratic forms is the assertion that 5 and 13 are ruled out.

Anyway, Sillke had also stated another conjecture, based on the observation that a prime p > 3 divides some integer s[n]-1 if and only if the period of the recurrence of s[j] (mod p) is a multiple of 6. This conjecture is not too difficult to prove. First, let's review a little what's known about the periods of s[j] modulo primes.

(We'll restrict the discussion to 2nd-order recurrences modulo rimes; for a disucssion of recurrence periods of general nth-order recurrences modulo primes and composites, see Section 2.7 of Symmetric Pseudoprimes.

We know the period of s[n] mod p divides p+1 if 3 is not a quadratic residue mod p, and divides p-1 otherwise. Thus, if we denote the period of s[n] mod p by T[p], we have

$$T[p] = (p+1)/k \quad \text{if } p = 5 \text{ or } 7 \pmod{12}$$

$$(p-1)/k \quad \text{if } p = 1 \text{ or } 11 \pmod{12}$$

Of course, the primes congruent to 5 or 7 (mod 12) are congruent to 5,7,17, or 19 (mod 24), and the primes congruent to 1 or 11 (mod 12) are congruent to 1,11,13, or 23 (mod 24).

Now, remember that the only primes that divide s[n]-1 are congruent to 1, 13, or 17 (mod 24). If p is congruent to 1 or 13 then it is of the form p=12j+1, and so

$$T[p] = \frac{p-1}{k} = \frac{(12j+1)-1}{k} = \frac{12j}{k}$$

On the other hand, if p is congruent to 17 (mod 24) then it is of the form p=12j+5 and we have

$$T[p] = \frac{p+1}{k} = \frac{(12j+5)+1}{k} = \frac{6(j+1)}{k}$$

This shows that if p' divides some s[n]-1 then T[p] must be a multiple of 6 *IF* k = 1, i.e., if the recurrence corresponds to a primitive root (mod p).

Conversely if p is congruent to 7 or 11 (mod 12), it cannot divide any s[n]-1, and T[p] cannot be divisible by 6, because in those cases the numerator of T[p] is either 12j+8 or 12j+10, neither of which is divisible by 6 for ANY value of k.

In view of all this, it's clear why Sillke's conjecture is likely to apply to most primes. The only uncertainty is whether k might divide out the 6 from the numerator of some period. This is essentially the same phenomenon that allows pseudoprimes

Some Properties of the Lucas Sequence

to occur, i.e., aliasing, so that a field with one characteristic "looks like" a field with another, because k > 1.

To illustrate, here's a list of the primes that are in the "right" congruence classes to be possible divisors of s[n]-1, along with their periods and the corresponding values of k. I'll put an asterisk (*) next to those that do not divide s[n]-1:

p	T[p]	k
13	12	1
17	18	1
37	36	1
41 *	14	3
61	60	1
73	36	2
89	90	1
97 *	16	6
109	108	1
113	114	1
137	138	1
157	156	1
181 *	20	9
193	24	8
229	228	1
233	234	1
241	30	8
257	258	1
277 *	92	3
281	282	1
313	78	4
337 *	56	6
349 *	116	3
353 *	118	3
373 *	124	3
397	396	1
401	402	1
409	204	2
421	420	1
433	216	2

449	150	3
457	228	2
521 *	58	9
541	540	1
569	114	5
577	288	2
593	594	1
601	300	2
613	204	3
617	618	1
641	642	1
661	60	11
673	336	2
709	708	1
733 *	244	3
757	84	9
761	762	1
769	192	4
809	810	1
829	276	3
853 *	284	3
857	858	1
877	876	1
881	294	3
929	930	1
937	234	4
953	954	1
977 *	326	3
997	996	1

As required, gcd(k,6) > 1 for all cases where T[p] is not divisible by 6. In fact, all the non-divisors have k a multiple of 3. This might lead us to suspect the converse, i.e., that p cannot be a divisor if k is a multiple of 3, but that's is certainly not the case, as shown by the primes 449, 613, 757, 829, etc. On the other hand, we might suspect that p MUST be a divisor if k is not divisible by 3, and this is certainly true for all p < 1000. This really justexpresses the fact that all the periods are even numbers, so divisibility by 3 automatically implies divisibility

Some Properties of the Lucas Sequence

By the way, notice that in most (2/3) of the cases, the numerator of T[p] is automatically not only a multiple of 6, it is a multiple of 12, so it has a power of 2 to spare. Also, in 1/3 of the cases the numerator is a multiple of 24, so it has TWO powers of 2 to spare.

Anyway, we're now in a position to prove Sillke's second conjecture, i.e., the observation that the period of the recurrence

$$s[0] = 2$$
$$s[1] = 4$$
$$s[n] = 4\ s[n-1] - s[n-2]$$

modulo p is divisible by 6 if any only if p divides s[n]-1 for some integer n. The key identity is

$$s[n+k] + s[n-k] = s[n]\ s[k] \qquad (1)$$

for any integers n and k. (Notice that with n=1 this is simply a re-statement of the basic recurrence, and it's easy to show by induction that it's true for all n.) Now, suppose a prime p divides s[n]-1 for some particular index n. In other words, we have s[n]=1 (mod p). It follows from (1) if we set k=n we have

$$s[2n] = s[n]^2 - 2 \qquad (2)$$

$$= (1)^2 - 2 \qquad (\text{mod } p)$$

$$= -1 \qquad (\text{mod } p)$$

(Of course, it also follows that s[4n], s[8n], etc., are all congruent to -1 (mod p).) We also know that s[k]=s[-k] for all k, so on the condition that there is an integer n such that s[n]=1 (mod p) we have

$$\begin{aligned} s[-2n] &= -1 \\ s[\ -n] &= 1 \\ s[\ \ 0] &= 2 \qquad (\text{mod } p) \\ s[\ \ n] &= 1 \\ s[\ 2n] &= -1 \end{aligned}$$

This provides more than enough values to infer the recurrence relation for the sequence in steps of n:

$$s[n(k)] = s[n(k-1)] - s[n(k-2)] \pmod{p} \qquad (3)$$

so we have the complete cycle

$$\begin{aligned} s[\,0] &= 2 \\ s[\,n] &= 1 \\ s[2n] &= -1 \\ s[3n] &= -2 \\ s[4n] &= -1 \\ s[5n] &= 1 \\ s[6n] &= 2 \end{aligned} \pmod{p}$$

which repeats ad infinitum. The period of the recurrence (mod p) must be a multiple of this cycle, so it must be divisible by 6. So we've proven that if p divides some s[n]-1 then the period of the recurrence (mod p) must be divisible by 6.

What about the converse? If the period of the recurrence is a multiple of 6, must we necessarily find that s[n]=1 (mod p) for some index n? Yes, because if the period of the recurrence is 6m we know that s[6m]=2 (mod p), which by equation (2) implies that s[3m] = +2 or -2 (mod p). We also have the relation

$$2\,s[3m] = 3\,s[m]\,s[2m] - s[m]^3$$

which implies that

$$s[m] = \begin{cases} +2 \text{ or } -1 & \text{if } s[3m] = +2 \\ -2 \text{ or } +1 & \text{if } s[3m] = -2 \end{cases}$$

Now, it's important to realize that we can rule out any value of the sequence being congruent to +2 (mod p) between s[0] and s[6m], because the sequence has a fundamental period j if and only if j is the least positive integer such that s[j]=2. This is due to the fact that the meta-sequence s[-j], s[0], s[j], s[2j],... has a period of 1, and therefore every sequence s[kj + r] has period 1. 'A full account of this is given in Section 2.5 of Symmetric

Some Properties of the Lucas Sequence

Pseudoprimes. The key point is that the recurrence for the sequence s[kn] is

$$s[kn] = s[k] \, s[k(n-1)] - s[k(n-2)]$$

and so if s[k]=2 (mod p) for any index k we have

$$s[kn] = 2 \, s[k(n-1)] - s[k(n-2)]$$

which has the characteristic polynomial

$$x^2 - 2x + 1 = (x-1)^2$$

In other words, it factors into the degenerate 1st-order recurrence s[kn] = s[k(n-1)], and since n can be any rational number such that kn is an integer, we can set n = m + r/k to give

$$s[km+r] = s[k(m-1)+r]$$

which shows that the entire s sequence has period k.

Therefore, since the premise is that 6m is the fundamental period of the recurrence (mod p), we know that s[3m] = -2, which implies that s[m] = +1 or -2. However, if s[m]=-2 then s[2m]=+2, which is impossible, so we must have s[m]=1 (mod p), which proves that p divides s[m]-1. This concludes the proof of Sillke's 2nd conjecture.

By the way, it's interesting to observe that if p divides some s[n]-1 then we have the meta sequence of values

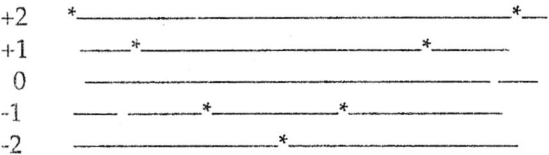

Moreover, recall from my previous email that most of the periods are actually divisible by 12, and some by 24, and in those cases we know that p divides s[u] itself, which means that s[u]=0

(mod p) for some u. Indeed, we find that the 0's always occur mid-way between the +1 and -1, so we have

```
+2    *_____*___
+1    ____*_____*_____
 0    _____*_____*_____
-1    _____*_____*_____
-2    _____*_____
```

These are precisely the lattice values of 2cos(2pi k/T) where T is the period of the recurrence modulo p. This function will give a value of 1 if and only if there is an integer k such that k/T = 1/6, which there will be iff the period T is a multiple of 6.

It isn't surprising that we have this relation, because notice that the recurrence of the meta-sequence S[k]=s[nk] given by equation (3) has the characteristic equation x^2 - x + 1, whose roots are the primitive cube roots of -1, i.e.,

r1 = (1+sqrt(-3))/2   r2 = (1-sqrt(-3))/2

and of course the values of the recurrence are simply

S[k] = (r1)^k + (r2)^k

If we write r1 and r2 in exponential form we have

r1 = e^(ia)   r2 = e^(-ia)

Recall that the definition of the cosine function is

$$\cos(q) = \frac{e^{(iq)} - e^{(-iq)}}{2}$$

so we have S[k] = 2 cos(k pi/6).

17

On a Unit Fraction Question of Erdos and Graham

Guy's excellent book "Unsolved Problems in Number Theory" (2nd Ed) it discusses expressing 1 as the sum of t distinct unit fractions. Letting m(t) denote the smallest possible maximum denominator in such a sum, he notes that m(3)=6 and m(4)=12. These follow from the optimum 3-term and 4-term expressions

$$1 = 1/2 + 1/3 + 1/6$$

$$1 = 1/2 + 1/4 + 1/6 + 1/12.$$

However, the book goes on to say that m(12)=120. Is this just a typo? If I've interpreted the definition of m(t) correctly it seems to me m(12) cannot be greater than 30. Here's a table of the optimum expansions for t=3 to 12:

t	denominators of optimum expansion						
3	2	3	6				
4	2	4	6	12			
5	2	4	10	12	15		
6	3	4	6	10	12	15	
7	3	4	9	10	12	15	18

t	denominators of optimum expansion									
8	3	5	9	10	12	15	18	20		
9	4	5	8	9	10	15	18	20	24	
10	5	6	8	9	10	12	15	18	20	24
11	5	6	8	9	10	15	18	20	21	24	28
12	6	7	8	9 10 14	15	18	20	24	28	30

I'm not actually certain the above expansion for t=12 is optimum, but it proves that the maximum denominator of the optiumum expansion is certainly no greater than 30. Am I missing something?

18

The Greedy Algorithm for Unit Fractions

Suppose we want to write the simple fraction 2/3 as a sum of unit fractions with distinct odd denominators. If we apply the "greedy algorithm", which consists of taking the largest qualifying unit fraction at each stage, we would begin with the term 1/3, leaving a remainder of 1/3. Since we require distinct denominators we can't use 1/3 for our second term, so we go on to the next largest odd unit fraction, which is 1/5. This leaves a remainder of 2/15, and the largest odd unit fraction less than 2/15 is 1/9, so that is our third term, leaving a remainder of exactly 1/45.

So, the odd greedy expansion of 2/3 terminates after four steps, giving the result

$$2/3 = 1/3 + 1/5 + 1/9 + 1/45$$

The non-zero remainders we encountered during this process were 1/3, 2/15, and 1/45, with the numerators 1, 2, 1.

There are some interesting unsolved problems related to odd greedy expansions. One open question is whether every fraction is guaranteed to terminate in a finite number of steps. This is not a trivial question, as shown by the odd greedy expansion of 1 (not using 1/1 on the first step). The result is

1 = 1/3 + 1/5 + 1/7 + 1/9 + 1/11 + 1/13 + 1/23 + 1/721 + 1/979007

+ 1/661211444787 + ...

It isn't known whether this eventually terminates. Incidentally, if we restrict ourselves to prime denominators we have

1 = 1/2 + 1/3 + 1/7 + 1/43 + 1/1811 + 1/654149 + ...

Most fractions terminate quickly under the application of the odd greedy algorithm, but some require quite a large number of terms. For example, if the fraction 3/179 is converted into a sum of unit fractions with odd denominators using the greedy algorithm, it takes 19 terms. Also, the numerators of the sequence of remainders is

3, 4, 5, 6, 7, 8, 9, 10, 11, 12, 13, 14, 15, 16, 17, 2, 3, 4, 1

Of course, if we don't require the use of the greedy algorithm, then the fraction 3/179 has the 3-term expansion with odd denominators 1/63 + 1/1611 + 1/3759, and conversely if allow both odd and even denominators then the greedy algorithm yields the 2-term expansion 1/60 + 1/10740. It is only the simultaneous imposition of BOTH requirements (odd AND greedy) that leads to the unusually lengthy expansion. Note also that the numerators of successive terms are consecutive integers from 3 to 17. (This raises the question of whether you could "construct" examples by working backwards from a given sequence of remainders.)

In general, if we have a fraction N/D we can generate an expansion

$$\frac{N}{D} = \frac{1}{d[0]} + \frac{1}{d[1]} + \frac{1}{d[2]} + \frac{1}{d[3]} + ...$$

that is quadratically convergent (i.e., the number of correct digits roughly doubles with each term) using the recurrence

$$(N+k)\, d[k+1] = (N+k-1)\, d[k]^2 - (N+k)\, d[k] + (N+k+1)$$

The Greedy Algorithm for Unit Fractions

with the initial value $d[0] = 1 + (D+1)/N$. We can re-write this recurrence in the form

$$d[k+1] = d[k]^2 - d[k] + 1 - \frac{d[k]^2 - 1}{N+k} \quad (1)$$

Of course this doesn't guarantee that the values of $d[j]$ are necessarily integers. To make $d[0]$ an integer we must have $D \equiv -1 \pmod{N}$. Thereafter on the kth step we must have $d[k]^2 - 1$ divisible by $(N+k)$, which implies that $d[k] \equiv +1$ or $-1 \pmod{N+k}$. Taking the fraction $5/179$ as an example, we have $N=5$ and $D=179$, which gives

$d[0] = 37$
$d[1] = 1105$
$d[2] = 1045489$
$d[3] = 956415297493$
$d[4] = 813093530024486866555885$
$d[5] = 297504448985545650649017654567005656145138938200 93/5$

This gives a unit fraction expansion up until the denominator $d[5]$, which is not an integer because $d[4] \equiv 4 \pmod 9$, so it's not congruent to $+1$ or $-1 \pmod{N+k}$. As a result, the sequence of remainders is

$$6, 7, 8, 9, 2, \ldots$$

The numerator of

$$N/D - 1/d[0] - 1/d[1] - 1/d[2] - 1/d[3] - 1/d[4]$$

should be 10, but $d[4]$ happens to be divisible by 5, so the reduced numerator is 2. As a result the recurrence stops giving integers, because it's based on the assumption that the remainders increase by 1 at each step. Of course, we can start the recurrence over again with this new value of N/D.

For another example, consider again the fraction $3/179$. In this case the recurrence formula (1) gives

$$d[0] = 61$$
$$d[1] = 2731$$
$$d[2] = 5963959$$
$$d[3] = 29640666497443$$
$$d[4] = 753059237496518829212535343$$
$$d[5] = 496210938281483556785833636950652507016084391058576351$$

and so on

Recurrence (1) with the initial value $d[0]=61$ gives integer values for a long time, up to 19 terms. The factorizations of $d[k]-1$ are somewhat cumulative, as shown below

$$d[0]-1 = (2)(2)(3) \quad (5)$$
$$d[1]-1 = (2) \quad (3) \quad (5)(7) \quad (13)$$
$$d[2]-1 = (2) \quad (3)(3) \quad (7)(11)(13)(331)$$
$$d[3]-1 = (2) \quad (3) \quad (7) \quad (13)(331) \quad (14909897)$$
$$d[4]-1 = (2) \quad (3) \quad (11)(13)(331)(7505)(77839)(323759)(14909897)$$
$$d[5]-1 = (d[4]-1)(3)(3)(5)(5)(\text{big composite})$$

If we let $f\_N(n)$ denote one less than the index of the first non-integer value given by the recurrence formula (1) with the initial value $d[0]=n$, then we saw previously that $f\_5(37) = 4$. I've also found that $f\_3(51)=2$, and evidently $f\_3(61)=13$. Of course, this doesn't completly describe an expansion, because once the integers break down we can start over again with the new N/D, as illustrated by the case 3/179:

```
3, 4, 5, 6, 7, 8, 9, 10, 11, 12, 13, 14, 15, 16, 17,    2, 3, 4,    1
|————————————————————————————————————|  |————|  |-|
```

By the way, there may be some connection with a sequence studied by Sylvester, defined in Sloane's Handbook of Integer Sequences (M0865) as

$$a(n+1) = a(n)^2 - a(n) + 1$$

This is the same as formula (1) except it doesn't have the last term involving N.

We might try to estimate, on a naive probabilistic basis, the

The Greedy Algorithm for Unit Fractions

expected value of f_N(n) for any given N and n, because it seems to depend only on whether each d[k] is or is not congruent to +-1 (mod N+k). If d[k] were equally likely to be in any of the N+k equivalence classes mod N+k, this would give a probability of 2/(N+k) that the kth step will give an integer. Thus we might infer that small numerators N will give the best chance for long strings, and the probability of j consecutive integers would be something like

$$\frac{N!}{(N+j)!} \, 2^{\wedge}j$$

However, if this were correct, the case 3/179 would have a probability of about 1 in 425,675,250. Evidently the assumption of equi-probable equivalence classes is wrong.

Here's a short of list of fractions that act somewhat like 3/179 under iterations of

$$x[k] \;=\; x[k-1]^{\wedge}2 \;-\; x[k-1] \;+\; 1 \;-\; \frac{x[k-1]^{\wedge}2 - 1}{N+k} \qquad (1)$$

Table of "Persistently Odd & Greedy" Denominators

			numerators			
3	5	7	11	13	17	19
179	139	12473	1627	50387\	218483\	168149
197	1399	18143	14299	84239/	223651/	223211
377	2209	20663	17071	129947	366283\	334931
629	2699	22049	41381	260597	371449/	
827	3649	24023	46331	594047\	398071	
1079	4909	32129	58057	627899/	589933	
1367	5039	40193	77417	674309		
1619	5809	43679	99439			
1817	5849	45863				
1997	7289	46073				
2069	8549					
2249						

For example, the first "stubborn" fraction with numerator 5 is 5/139.

The next is 5/1399, and so on. Each of these gives integer values for AT LEAST the initial 9 steps of the recurrence. To go beyond this, we can evaluates the recurrence (mod p) for various primes p, which enables us to determine how long the sequence goes on giving integers (although it won't tell what those integers are).

Obviously every one of these stubborn fractions N/D has D=-1 (mod 2N).

In addition, it seems that if N = 1 (or 0) (mod 3) then D=-1 (mod 6N).

I didn't include N=1 in the above table, but it's interesting to determine the persistent denominators for that case as well. It seems that the most persistent are D = 19, 61, 73, 151, 181, 193, 271, 283, 379, and so on.

The first 10 denominators of the odd-greedy expansion for

$$5/139 = 1/29 + 1/673 + ...\text{etc.}$$

are given here. The sequence terminates after 19 steps, and the numerators of the remainders are

5,6,7,8,9,10,11,12,13,14,15,16,17,26,51,2,3,4,1

We would like to efficiently determine an upper limit for the number of consecutive iterations of (1) that could possibly yield integers, beginning with the initial value $x[0] = (D+1)/N + 1$ for any given fraction N/D.

The only simple way of doing this that I can see is by evaluating the iteration (mod p) for various primes p > N+1. If $x[p-N-1]$ is not congruent to +1 or -1 (mod p), then the string must be broken at (or before) the point when N+k reaches p.

The limiting primes by which point the string must break down for various fractions are listed below

3/179	17	5/139	17
3/197	17	5/1399	19
3/377	13	5/2209	19

The Greedy Algorithm for Unit Fractions 145

3/629	13		5/2699	17
3/827	13		5/3649	17
3/1079	13		5/4909	17
3/1367	41		5/5039	17
3/1619	17		5/5809	31
3/1817	19		5/5849	17
3/1997	13		5/7289	17
3/2069	29		5/8549	17
3/2249	13			
3/2267	13			
3/2447	19			
3/2699	13			
3/2879	23			
3/2897	13			
3/2969	13			

This suggests that 3/1367, 3/2069, 3/2879, and 5/5809 would be good candidate for highly "stubborn" odd greedy expansions, since the sequences of uniformly increasing remainders in these cases don't *necessarily* break down until N+k equals 41, 29, 23, and 31 respectively.

Of course, they COULD break down much sooner. The others (the 13's and 17's) definitely won't yield expansions of length greater than 17.

Another interesting approach to this problem is to determine the sufficient condition for greediness of a sequence. Clearly if 1/n is a term in the sequence, then the sum of all later terms in the sequence must be less than the difference

$$1/n - 1/(n+2) = ((n+2)-(n))/(n(n+2)) = 2/(n(n+2))$$

Now, for all odd n (greater than 1) we have the inequality

$$2/(n(n+2)) > 1/n^2 + 1/n^4 + 1/n^8 + \ldots$$

from which it follows that if each denominator is more than the square of the preceeding one, then the sequence is greedy. The question is whether the sum of an infinite greedy sequence can ever be rational. (Of course, I haven't shown that this is a *sufficient condition* for greed, but merely a necessary condition. Thus, there could be greedy sequences that do not satisfy the "quadratic condition".) This question is answered in Irrationality of Quadratic Sums. Also, a method of constructing fractions with arbitrarily long odd greedy expansions with consecutive integer remainder numerators is presented in Odd-Greedy Unit Fraction Expansions.

19

Average of Sigma(n)/n

Hardy and Wright's "Introduction to the Theory of Numbers" gives a simple demonstration that the average "order" of sigma(n)/n is equal to zeta(2) (the sum of the inverse squares) based a count of the lattice points in a plane region, but there are some other interesting approaches to this problem. For example, by the Taylor series expansion of the natural log function, we have

$$\ln(1-x) = -\left(\frac{x}{1} + \frac{x^2}{2} + \frac{x^3}{3} + \frac{x^4}{4} + \ldots \right) \quad (1)$$

for all $|x|$ less than 1. Now, if we consider the sum of the natural logs of $1-x$, $1-x^2$, $1-x^3$, and so on, we have

$$-\sum_{n=1}^{\infty} \ln(1 - x^n) = \frac{x}{1} + \frac{x^2}{2} + \frac{x^3}{3} + \frac{x^4}{4} + \frac{x^5}{5} + \frac{x^6}{6} + \ldots$$

$$+ \frac{x^2}{1} + \frac{x^4}{2} + \frac{x^6}{3} + \frac{x^8}{4} + \frac{x^{10}}{5} + \frac{x^{12}}{6} + \ldots$$

$$+ \frac{x^3}{1} + \frac{x^6}{2} + \frac{x^9}{3} + \frac{x^{12}}{4} + \frac{x^{15}}{5} + \frac{x^{18}}{6} + \ldots$$

$$+ \frac{x^4}{1} + \frac{x^8}{2} + \frac{x^{12}}{3} + \frac{x^{16}}{4} + \frac{x^{20}}{5} + \frac{x^{24}}{6} + \ldots$$
$$+ \text{etc}\ldots$$

Combining terms by powers of x gives the nice identity

$$-\sum_{n=1}^{\infty} \ln(1 - x^n) = \sum_{n=1}^{\infty} \frac{\sigma(n)}{n} x^n \qquad (2)$$

If we divide both sides of this relation by the geometric series, i.e., the sum of x^n for n=1 to oo, we have

$$-\frac{1-x}{x} \sum_{n=1}^{\infty} \ln(1 - x^n) = \frac{\sum_{n=1}^{\infty} \frac{\sigma(n)}{n} x^n}{\sum_{n=1}^{\infty} x^n} \qquad (3)$$

Notice that the right side of this equation a geometrically weighted average of all the values of sigma(n)/n from n=1 to infinity. Also, as x approaches 1, this approaches the EVENLY weighted average out to arbitrarily large n. This suggests that it would be interesting to evaluate the limit of the left side as x goes to 1. This turns out to be sort of a delicate operation, because if the summation on the left side of (3) is evaluated for n=1 to N for any fixed N, the overall quantity on on the left side goes to zero as x goes to 1. However, for any fixed x less than 1, the quantity converges on a non-zero value as N goes to infinity, and these limiting values converge on a certain value as x goes to 1.

Anyway, by making x sufficiently close to 1 we can make the ratios of successive terms in the left-hand summation arbitrarily close to 1, so the summation can be approached by the integral in the limit as x goes to 1. In other words

Average of Sigma(n)/n 149

$$\lim_{x \to 1} \sum_{n=1}^{\infty} \ln(1-x^n) = \int_{n=1}^{\infty} \ln(1-x^n)\, dn \qquad (4)$$

To evaluate the integral, let's make the substitution $q = x^n$, where q goes from x to 0 as n goes from 1 to infinity. Noting that q can be written as $\exp(n\ln(x))$ we have

$$\frac{dq}{dn} = \exp(n\ln(x))\ln(x) = \ln(x)\, x^n \qquad (5)$$

which gives

$$dn = \frac{1}{\ln(x)\, x^n}\, dq = \frac{1}{q\ln(x)}\, dq$$

With these substitutions the integral becomes

$$\int_{n=1}^{\infty} \ln(1-x^n)\, dn = \frac{1}{\ln(x)} \int_{q=x}^{0} \frac{\ln(1-q)}{q}\, dq \qquad (6)$$

Recalling the power series expansion of $\ln(1-q)$, we have the nice indefinite integral

$$\int \frac{\ln(1-q)}{q}\, dq = -\left(\frac{q}{1^2} + \frac{q^2}{2^2} + \frac{q^3}{3^2} + \frac{q^4}{4^2} + \ldots \right) \qquad (7)$$

Evaluating this from $q=x$ to $q=0$ gives

$$\int_{q=x}^{0} \frac{\ln(1-q)}{q}\, dq = \frac{x}{1^2} + \frac{x^2}{2^2} + \frac{x^3}{3^2} + \frac{x^4}{4^2} + \ldots$$

Substituting this back into equation (6), and from there back into equation (3), we arrive at the expression for the evenly weighted average of sigma(n)/n for all integers n:

$$\lim_{x \to 1} -\frac{1-x}{x \ln(x)} \left(\frac{x}{1^2} + \frac{x^2}{2^2} + \frac{x^3}{3^2} + \frac{x^4}{4^2} + \ldots \right)$$

Since the ratio of $-(1-x)$ to $\ln(x)$ goes to 1 as x approaches 1, and the numerators inside the parentheses also go to 1, we have the result

$$\lim_{x \to 1} \frac{\sum_{n=1}^{\infty} \frac{\sigma(n)}{n} x^n}{\sum_{n=1}^{\infty} x^n} = \sum_{n=1}^{\infty} \frac{1}{n^2}$$

which shows, as expected, that the average value of sigma(n)/n for all integers is zeta(2) = pi^2 / 6. This approach is certainly much less economical than the simple lattice-point derivation presented in Hardy and Wright, but it does provide the opportunity to relate some common power series expansions to arithmetic functions. For example, from the expansions of (1 - 1/x)ln(1-x^n) we see that the ratio sigma(n)/n equals the sum of the numbers in the first n columns of the following array:

Average of Sigma(n)/n

```
1  -1/2  -1/6  -1/12  -1/20  -1/30  -1/42  -1/56  -1/72  -1/90  -1/110
    1    -1    1/2   -1/2   1/3   -1/3   1/4    -1/4   1/5    -1/5
          1    -1    1/2   -1/2                  1/3   -1/3
                1    -1                   1/2   -1/2
                      1    -1                          1/2    -1/2
                            1    -1
                                  1    -1
                                        1    -1
                                              1    -1
                                                    1    -1
                                                          1
```

This is related to the interesting fact that the series expansion of the left side of (3) is

$$1 + \frac{1}{2}x - \frac{1}{6}x^2 + \frac{5}{12}x^3 - \frac{11}{20}x^4 + \frac{24}{30}x^5 - \dots$$

where the coefficient of x^k is sigma(k)/k - sigma(k-1)/(k-1). This converges for all x less than 1, whereas for x=1 the partial sums are just the successive values of sigma(k)/k, so it never converges. Nevertheless, the limit of the convergent values as x approaches 1 is (pi^2)/6.

20

Lucas's Primality Test With Factored N-1

Fermat's Little Theorem assures us that if N is a prime then

$$b^{(N-1)} = 1 \pmod{N} \qquad (1)$$

for every integer b coprime to N. In contrast, if N is composite it is quite rare for the above congruence to be satisfied for ANY b. This fact enables us to test for compositeness quite easily, simply by checking to see if $b^{(N-1)}$ is congruent to 1 (mod N) for some particular value of b. However, although this congruence is necessary for primality, it isn't quite sufficient, because for any given base b there exist composites N that satisfy (1). Such integers are called pseudoprimes to the base b. In fact, there exist composite integers that are pseudoprimes to every base with which they share no common factor. Such composites are called Carmichael numbers, the smallest of which is 561.

Euler generalized Fermat's theorem by showing that for ANY integer N, prime or composite, we have

$$b^{\phi(N)} = 1 \pmod{N} \qquad (2)$$

for any integer b coprime to N, where phi(N) is Euler's totient function, representing the number of positive integers less than

Lucas's Primality Test With Factored N-1

and coprime to N. Obviously if N is a prime then phi(N) = N-1, in which case Euler's theorem reduces to Fermat's.

Edouard Lucas noticed that Euler's theorem enables us to definitely determine the primality of N fairly easily if we know the factorization of N-1. The basic idea of this test is the foundation for virtually all deterministic primality tests, so it's worthwhile to understand exactly how it works.

Suppose we examine the residues (mod N) of the sequence

$$1 \quad b \quad b^2 \quad b^3 \quad b^4 \quad b^5 \ldots.$$

Let T denote the smallest exponent such that $b^T = 1$ (mod N). This implies that b^k can be congruent to 1 (mod N) only if k is a multiple of T. We can call T the fundamental period of the sequence modulo N.

Also, from Euler's theorem we know that $b^{phi(N)} = 1$ (mod N), so clearly phi(N) is a multiple of T. Thus we can write phi(N) = mT for some integer m.

Now, suppose we have determined by calculation that the number N satisfies Fermat's condition for primality with respect to the base b. In other words, we have found that

$$b^{(N-1)} = 1 \quad (\text{mod } N)$$

It follows that N-1 must be a multiple of the fundamental period T, so we can write N-1 = nT for some integer n. Thus we know that the fundamental period T divides both phi(N) and (N-1). Of course, we also know that phi(N) is less than or equal to N-1. Therefore, if we could prove that the fundamental period T equals N-1, then it would follow that phi(N) equals N-1, which can only be the case if N is a prime.

But how can we prove that T = N-1? Well, we know that N-1 = nT for some integer n, which means that n divides N-1. Assuming n is greater than 1, let p denote any prime divisor of n. It follows that p is also a prime divisor of N-1, and we must have

$$b^{((N-1)/p)} = 1 \quad (\text{mod } N)$$

On the other hand, suppose we examine $b^{((N-1)/q)}$ for EVERY prime divisor q of N-1 and we find that none of those values is congruent to 1 (mod N). This can only mean that n=1, which means the fundamental period T equals N-1, and therefore phi(N) equals N-1 (because T divides phi(N), and phi(N) is less than or equal to N-1).

This proves Lucas's primality criterion:

If, for some integer b, the quantity $b^{(N-1)}$ is congruent to 1 modulo N, and if $b^{((N-1)/q)}$ is NOT congruent to 1 modulo N for ANY prime divisor q of N-1, then N is a prime.

Of course, this all just amounts to the assertion that if we can find an element b whose "order" mod N is N-1, then N is a prime. Such a number is called a "primitive root" mod N. If N is a prime there are phi(N-1) primitive roots modulo N, so we are assured of the existence of this many values of b that would be suitable to prove the primality of N via Lucas's test.

21

One In The Chamber

A well-known puzzle, attributed to Lucas, involves stacking cannon balls in a square-based pyramid. The apex has 1^2 ball(s), the second layer down has 2^2 balls, the third layer 3^2 balls, and so on. The classic question asks "How many layers are necessary so that the total number of balls is a perfect square?" What if we alter this question to ask for the number of layers necessary to give a perfect cube? This might seem more natural, assuming the balls come packed in cubical crates.

The sum of the first k squares is $(k)(k+1)(2k+1)/6$, so if this is to equal the cube of some integer N we have

$$(k)(k+1)(2k+1) = 6N^3$$

Clearly the factors k, k+1, and 2k+1 are mutually coprime. In view of divisibility requirements modulo 9, this equation implies the existence of integers a,b,c such that one of the following four sets of conditions apply:

$k = 6a^3$ $k+1 = b^3$ $2k+1 = c^3$

or

$k = a^3$ $k+1 = 6b^3$ $2k+1 = c^3$

or

$k = 2a^3$ $k+1 = b^3$ $2k+1 = 3c^3$

or

$k = a^3$ $k+1 = 2b^3$ $2k+1 = 3c^3$

Therefore, since $(k) + (k+1) + (-(2k+1)) = 0$, we're led to an equation that is either of the form

$$x^3 + y^3 + 6z^3 = 0$$

or else

$$x^3 + 2y^3 + 3z^3 = 0$$

These equations are closely related to each other and, in general, solutions to them are quite rare, even without the additional requirements that the components be of the form k, $k+1$, and $2k+1$.

In fact, it seems that Legendre once (mistakenly) stated there were no integer solutions of $x^3 + y^3 = 6z^3$. For more on this problem, and what Legendre might have been thinking, see the article The 450 Pound Problem.

The above conditions preclude the existence of any solution other than the trivial ones. This follows because the four possibilities listed above lead, respectively, to

$$c^3 + (+1)^3 = 2b^3$$

or

$$c^3 + (-1)^3 = 2a^3$$

or

$$b^3 + (-1)^3 = 2a^3$$

or

$$a^3 + (+1)^3 = 2b^3$$

and, as Euler proved, the only integer solutions of $x^3 + y^3 = 2z^3$ are with $x=y$. Thus the only integers solutions of $k(k+1)(2k+1)=6N^3$ are $[k=1,N=1]$, $[k=0,N=0]$, $[k=-1,N=0]$, and $[k=-2,N=-1]$.

Incidentally, the classical problem of stacking the balls in a square-based pyramid has essentially the same answer as stacking the balls in a triangular-based pyramid, except multiplied by a factor.

Having shown there is no solution to the original problem, we might change the problem slightly by saying that, when unpacking the box we'll stack the rounds in a square pyramid, but we will place one round into the barrel. Thus we need a cube that equals a pyramid PLUS 1. If we ship the rounds in a cubical box of $26^3 = 17576$, we can put one round in the chamber and stack the other 17575 in a square-based pyramid of 37 layers.

22

Fractions and Characteristic Recurrences

It's easy to find the best fractional approximations for the square root of 2, based on the simple continued fraction. This gives convergents 7/5, 17/12, 41/29, and so on. However, it's not so easy to define the analagous sequence for CUBE root of 2.

Lagrange proved the simple continued fraction of an irrational number is periodic if and only if the number is a quadratic irrational, i.e., the root of a quadratic equation, such as $x^2 - 2 = 0$. As a result, the numerators and denominators of successive convergents of the continued fraction for sqrt(N) can always be generated by a simple second order linear recurrence.

This corresponds to the solutions of the basic "Pell equation" $x^2 - Ny^2 = 1$. On the other hand, solutions of the analagous equation $x^3 - Ny^3 = 1$ for cubes are less well-behaved.

Nevertheless, it's always possible to construct a linear recurring sequence of integers s[0], s[1], s[2],... such that any specified algebraic number is approached by some function of the ratio of successive terms s[n+1]/s[n] In particular, to construct a sequence for the rth root of N, let m denote the largest integer such that $m^r < N$. Then we have the rth order recurrence relation

$$s[n] = \sum_{j=1}^{r} \binom{r}{j} m^{(r-j)} (N-m^r)^{(j-1)} s[n-j]$$

with the initial values s[0]=s[1]=..s[r-2]=0 and s[r-1]=1, any by construction we have

$$N^{\wedge}(1/r) = \lim_{n->inf} m + (N-m^{\wedge}r) \frac{s[n]}{s[n+1]}$$

For example, to construct the sequence for the cube root of 2 we have N=2 and r=3, so m=1 and the recurrence is
s[n] = 3 s[n-1] + 3 s[n-2] + s[n-3]

The approximations for $2^{\wedge}(1/3)$ given by successive ratios of terms in this sequence are listed below, along side the convergents of the continued fraction

generated by linear recurrence	convergents of continued fraction	1/ IN(N-DV) I
4/3 *	4/3	1.135
5/4 *	5/4	5.040
29/23 *	29/23	1.581
223/177	34/27	1.646
	63/50	4.021
286/227 *	286/227	1.682
3301/2620	349/277	1.533
635/504 *	635/504	7.530
5429/4309 *	5429/4309	0.940
6064/4813 *	6064/4813	12.546
723235/574032	90325/71691	0.924
463753/368081	96389/76504	8.964
10705243/8496757	1054215/836731	2.298
957826/760227	2204819/1749966	1.368
52819267/41922680	3259034/2586697	3.787
15240955/12096754 *	15240955/12096754	8.806
2345474521/186104361	186150494/147747745	1.834

Those marked with an asterisk are the same on both lists. This raises the interesting question of whether infinitely many convergents occur in the sequence generated by the 3rd order recurrence.

We might also wonder which of the above approximations is the "best". Obviously the absolute accuracy improves indefinitely, but we can define a notion of "goodness" by considering the ratio of how many digits of accuracy are achieved per digit of the numerator and denominator. One such measure for the approximation N/D to the value V is $1/|N(N-DV)|$. The goodness of each convergent is listed in the right hand column of the table above.

It's interesting that the goodness of the continued fractions never falls much below 1.0, whereas the linear recurrence approximations can have very low values of "goodness". For example, the goodness of 723235/574032 is just 0.0122.

I wonder what can be said about the distribution of "goodness" of continued fraction convergents. For example, how soon would we expect a convergent with more goodness than 12.546, which is the maximum for all the values listed in the table?

23

Automedian Triangles and Magic Squares

Given any triangle with sides of lengths 2A, 2B, 2C, the lengths of the medians to the midpoints of those sides, respectively, are

$$(M\_A)^2 = -A^2 + 2B^2 + 2C^2$$

$$(M\_B)^2 = 2A^2 - B^2 + 2C^2$$

$$(M\_C)^2 = 2A^2 + 2B^2 - C^2$$

An "automedian triangle" is defined as one whose three medians are proportional to the three sides. Obviously the middle edge (in terms of length) must map to the middle-sized median, whereas the largest edge must map to the smallest median, and vice versa.

Thus we have three equations of the form

$$3C^2 = -A^2 + 2B^2 + 2C^2$$

$$3B^2 = 2A^2 - B^2 + 2C^2 \qquad (1)$$

$$3A^2 = 2A^2 + 2B^2 - C^2$$

Each of these equations reduces to simply $A^2 + C^2 = 2B^2$, so any solution of this equation in integers gives an

Automedian Triangles and Magic Squares 161

automedian triangle with integer sides.

Of course, if we want a strictly real triangle we also have to impose the requirement that all three lengths satisfy the triangle inequality. For example, the smallest integer solution of (1) is {A,B,C} = {1,5,7}, which doesn't make a real triangle, but there are infinitely many others, such as {7,13,17}, {17,25,31}, {31,41,49}, {23,37,47}, etc. that do give real automedian triangles with integer sides.

Equations (1) have a very interesting "duality", which can be seen if we allow the variables on the left side to be any three quantities (not necessarily some permutation of A,B,C). In other words, we have the equations

$$3I^2 = -A^2 + 2B^2 + 2C^2$$

$$3H^2 = 2A^2 - B^2 + 2C^2 \qquad (2)$$

$$3G^2 = 2A^2 + 2B^2 - C^2$$

If we solve these three linear equations for the squares of A,B,C in terms of the squares of I,H,G we get

$$3A^2 = -I^2 + 2H^2 + 2G^2$$

$$3B^2 = 2I^2 - H^2 + 2G^2 \qquad (3)$$

$$3C^2 = 2I^2 + 2H^2 - G^2$$

which is formally identical to (2). Thus, the triples {A,B,C} and {I,H,G} are duals of each other. Now, if we restrict the variables to integer values, it turns out that almost the only solutions of (2) give {I,H,G} as some permutation of {A,B,C}, so they are automedian triangles (with integer sides).

This raises the question of whether there exist pairs of *distinct* triples of integers {A,B,C} and {I,H,G} satisfying equations (2) and (3). These would correspond to integer triangles whose medians are proportional to *each others* sides. One reason this is an interesting question is that if no such pair of "amicable triangles" exists, it follows that there does not exist a 3x3 magic square of (distinct) square numbers, which is an open problem in number theory.

In case the connection isn't obvious, consider a 3x3 magic square consisting of distinct squares

$$A^2 \quad B^2 \quad C^2$$
$$D^2 \quad E^2 \quad F^2$$
$$G^2 \quad H^2 \quad I^2$$

By assumption the sum of each row, column, and main diagonal is the same, and it's easy to see that this sum must be $3E^2$. Therefore, we immediately have the following necessary conditions

$$\begin{aligned} A^2 + I^2 &= 2E^2 \\ B^2 + H^2 &= 2E^2 \\ C^2 + G^2 &= 2E^2 \\ A^2 + B^2 + C^2 &= 3E^2 \end{aligned} \qquad (4)$$

Subtracting twice the 4th equation from three times each of the previous three equations immediately gives equations (2). Obviously if {I,H,G} is a permutation of {A,B,C} the entries of the magic square are not distinct, so all the "self-dual" solutions can be ruled out, leaving us only with the question of whether there exist any "amicable" solutions in all distinct numbers.

It turns that "amicable" solutions exist, but they aren't very common. The first several primitive solutions are

A	B	C	I	H	G	E^2	factorization of E^2
49	421	541	559	371	149	157441	(29)(61)(89)
191	763	785	887	491	455	411625	(5^3)(37)(89)
361	941	1159	1201	829	479	786361	(37)(53)(401)
137	1123	1523	1543	1067	283	1199809	(13)(17)(61)(89)
931	1541	1691	1789	1301	1099	2033641	(41)(193)(257)
487	3647	4783	4903	3313	1183	12138289	(17^2)(97)(433)
53	4597	6223	6317	4333	1087	19953649	(61)(109)(3001)

A	B	C	I	H	G	E^2	factorization of E^2
5323	9005	10489	10861	8075	6023	73147825	(5^2)(73)(149)(269)
2447	11777	15487	15823	10847	4063	128177569	(13)(17^2)(109)(313)
10271	14939	16331	17071	13181	11411	198456241	(61)(1409)(2309)
8837	15467	16783	17923	12653	10847	199663249	(17)(73)(349)(461)
11215	16387	17041	18185	13709	12887	228235225	(5^2)(53)(281)(613)
6419	19283	19585	22133	12619	12145	265536625	(5^2)(53)(149)(269)
5071	20815	21997	24553	13975	12029	314282425	(5^2)(29)(41)(97)(109)
15953	23393	24553	26113	19727	18263	468193489	(89)(97)(193)(281)
3391	30085	30737	35063	18325	17209	620456425	(5^2)(53)(197)(2377)
7021	32653	39745	41803	27029	14735	898392625	(5^3)(13)(17^2)(1913)
2503	33827	41177	43487	27443	14207	948692089	(17)(149)(353)(1061)
11039	43727	51395	54727	34711	21805	1558452025	(5^2)(13)(29)(37)(41)(109)
17791	53675	61313	65737	41915	29641	2318936425	(5^2)(13)(17)(29)(41)(353)
34285	54173	61939	64205	48611	38227	2648871625	(5^3)(37)(41)(61)(229)
1523	53077	73693	74147	51797	8333	2750048569	(29)(37^2)(113)(613)
19177	56039	72755	74161	52223	23965	2933805625	(5^4)(1361)(3449)
9419	70651	74909	83899	46219	38941	3563879881	(17)(353)(401)(1481)
44405	61243	73751	73955	60749	44743	3720573025	(5^2)(13^2)(113)(7793)
22207	69395	77801	84151	52525	39007	3787270825	(5^2)(13)(149)(197)(397)
45653	68873	77593	80507	61823	50447	4282786729	(13)(29)(101)(137)(821)
17977	77635	84161	92911	54115	43273	4477813225	(5^2)(7561)(23689)
3001	67741	95041	95279	67069	7369	4543546921	(89)(173)(269)(1097)
42125	69283	86569	87205	67669	43417	4689613825	(5^2)(13)(41)(353)(997)
11659	80771	87871	97219	55349	43199	4793733121	(73)(113)(701)(829)
28729	72401	93881	95369	68401	33271	4960300801	(241)(2633)(7817)

41513 73195 90541 (5^3)(17)(61)(101)(389)	91991	69485	44587	5092836625
34391 77087 93395 (5^2)(37)(53)(277)(389)	96863	67991	42925	5282590825
27265 79717 94981 (5^2)(13^2)(521)(2441)	100015	66269	41533	5373190225
16885 84049 103343 (5^2)(17^3)(113)(433)	108325	70393	36601	6009704425
11095 83821 112447 (5^4)(17)(613)(1013)	114335	78547	23479	6597795625
20587 107309 110735 (5^2)(17^2)(41)(113)(241)	125341	67963	62225	8067095425
3167 107881 124295 (5^2)(13)(3181)(8737)	134369	80167	51145	9032529025
6889 110123 123635 (5^2)(13^2)(41)(53)(997)	135127	78611	54965	9153382225
30415 108683 128081 (5^2)(17)(29)(373)(2113)	136025	87269	54983	9713936425
82427 101483 111503 (13)(17^2)(37)(101)(701)	113533	96877	85153	9841976209
43409 107309 130151 (13)(37)(41)(193)(2657)	135431	93331	57329	10112948521
30395 119993 122599 (5^3)(13)(1933)(3221)	138965	76399	72143	10117563625
52303 116971 119125 (5^2)(73)(557)(10037)	132929	81997	78835	10202861425
57001 106171 133111 (29)(41)(101)(109)(821)	135071	101099	61439	10746644521
11095 119701 137407 (5^2)(17)(29)(53)(73)(233)	148655	88843	57799	11110704025
34577 132287 149177 (29)(73)(109)(149)(397)	161567	98993	71033	13649732209
34841 132685 171137 (5^2)(17^2)(1249)(1777)	175663	120275	52759	16035692425
7331 147239 171731 (73)(193)(509)(2381)	184651	111671	68251	17074867681
44375 164297 170621 (5^2)(17)(37)(157)(7841)	191695	108271	98003	19358056825
95851 157901 181969 (17)(53)(61)(173)(2357)	188771	141029	108209	22410952321
47905 196189 198623 (5^3)(13^2)(29)(149)(293)	226265	122477	118489	26745369625
49511 189449 224789 (109)(409)(557)(1193)	238321	152831	93371	29624119081
81121 200569 210541 (17)(41)(97)(613)(733)	232759	143281	128179	30378684361

Automedian Triangles and Magic Squares

(Some of these don't satisfy the triangle inequality, so they don't give "real" geometrical triangles.) Does this mean we have found 3x3 magic squares of squares? No, because we eliminated the central number E^2 from the magic square equations (4). Nothing we've done forces E^2 to be a square integer, and in fact E^2 is not a square for any of the amicable pairs of triangles that I've found.

In fact, it's clear that E^2 must be the product of at least three distinct primes of the form $4k+1$, and in order for E^2 to be a square it would be necessary for each of those primes to occur to an even power. It's possible that such cases don't exist, which is interesting because equations (4) represent only a subset of the conditions that would have to be met by a complete magic square of square integers.

For example, note that D^2 and F^2 don't even appear in those equations, and we haven't imposed the sums on the outer columns.

So, since no one has ever been able to find a 3x3 magic square of distinct square integers (nor prove that such is impossible), let me propose the simpler(?) problem of finding a 3x3 magic square of the following form

A^2	B^2	C^2
M	E^2	N
G^2	H^2	I^2

where M and N are integers but need not be squares. Thus I'm only requiring seven of the nine numbers to be squares, but they have to be the seven noted above, not just any seven, because it's known that there exists at least one magic square containing seven squares

23^2	205^2	289^2
373^2	425^2	222121
527^2	565^2	360721

but this doesn't satisfy the "amicable triangle" condition, i.e., it doesn't have both outer rows (or both outer columns) filled with squares, so it doesn't represent a solution of (4).

By the way, it's interesting that the "automedian" quality is really just an expression of eigenvectors and eigenvalues. In general, suppose we have the linear system

$$a11\ x1\ +\ a12\ x2\ +\ a13\ x3\ =\ y1$$
$$a21\ x1\ +\ a22\ x2\ +\ a23\ x3\ =\ y2 \qquad (5)$$
$$a31\ x1\ +\ a32\ x2\ +\ a33\ x3\ =\ y3$$

Now suppose we require the values of {y1,y2,y3} be proportional to the values {x3,x2,x1}. In other words, for some constant k we have

$$y1\ =\ k\ x3$$
$$y2\ =\ k\ x2$$
$$y3\ =\ k\ x1$$

Inserting these into (5) gives

$$a11\ x1\ +\ a12\ x2\ +\ (a13-k)\ x3\ =\ 0$$
$$a21\ x1\ +\ (a22-k)\ x2\ +\ a23\ x3\ =\ 0 \qquad (6)$$
$$(a31-k)\ x1\ +\ a32\ x2\ +\ a33\ x3\ =\ 0$$

Since the righthand vector is 0, the only non-trivial solutions require a value of k for which the determinant of the coefficient matrix is zero. Obviously the determinant of this matrix is a cubic in k, which has three roots, and those are called the eigenvalues of the system. By selecting any one of those eigenvalues, equations (6) typically give a solution [x1,x2,x3], which is called an eigen vector of the system.

Now, in the particular case of the automedian system we have

$$\begin{vmatrix} -1 & 2 & 2 \\ 2 & -1 & 2 \\ 2 & 2 & -1 \end{vmatrix} \begin{vmatrix} A^2 \\ B^2 \\ C^2 \end{vmatrix} = \begin{vmatrix} (Ma)^2 \\ (Mb)^2 \\ (Mc)^2 \end{vmatrix} \quad (3)$$

We want the medians to be proportional to the sides, so for some constant k we have

$$(Ma)^2 = k\, C^2 \qquad (Mb)^2 = k\, B^2 \qquad (Mc)^2 = k\, A^2$$

(Note that we've chosen to permute the vector components in these relations, but in general we could choose any permutation, including identity.) It follows that the value of k must be such that

$$\mathrm{Det} \begin{vmatrix} -1 & 2 & 2-k \\ 2 & -1-k & 2 \\ 2-k & 2 & -1 \end{vmatrix} = 0$$

Evaluating the determinant gives the "characteristic equation"

$$k^3 - 3k^2 - 9k + 27 = 0$$

which has the roots -3, 3, and 3. With k=3 the system degenerates to a single equation in three variables, $A^2 + C^2 = 2B^2$, corresponding to the automedian triangles. With k=-3 the system degenerates to two conditions, $A^2 = C^2$ and $B^2 + 2A^2 = 0$.

Obviously the only real solution of the second condition is A=B=0, but if we allow complex values we have the infinite family of solutions A=q, C=+-q, B=sqrt(-2)q.

24

Orthomagic Square of Squares

It's not known if there exists a 3x3 magic square of squares, i.e., a 3x3 arrangement of nine distinct integer squares such that the sum of each row, column, and main diagonal is the same. A recent note discussed one approach to this problem, namely, to determine the form of all 3x3 arrangements of squares that satisfy the four sums involving the central number, and then see if any of those arrangements can be made to also satisfy the four outer sums. In this way it was shown that no solution is possible if the central square is expressible as a sum of two squares in only four ways (which is the simplest non-trivial case). It may be possible to extend that method to the general case, but I wonder if another approach might be more effective.

Instead of looking at the 3x3 arrangements that satisfy the four sums involving the central number, suppose we consider the arrangements that satisfy the six orthogonal sums, i.e., the sums of the rows and columns. If these "orthomagic squares" of squares could be completely characterized, it might be possible to show that they can never satisfy the sums on the two main diagonals, thereby proving the impossibility of a 3x3 magic square of squares. (Of course, if this can't be shown, this approach may help to construct an example.)

Remarkably, it turns out that most orthomagic squares of squares also possess another property: the common sum of the rows and columns is a square! For example, the smallest orthomagic arrangement of distinct squares is

Orthomagic Square of Squares

$$4^2 \quad 23^2 \quad 52^2$$

$$32^2 \quad 44^2 \quad 17^2$$

$$47^2 \quad 28^2 \quad 16^2$$

and each rows and column of this arrangement sums to 3249 = 57^2. The same is true for the next several OMSOS's. In any case, this is nice because we know the common sum of a completely magic arrangement of squares must be of the form $3E^2$ where E^2 is the central square. Therefore, since a square can't be 3 times a square, we can immediately rule out all orthomagic arrangements whose common sum is a square.

Of the twelve smallest OMSOS's, nine of them have a square common sum, so this just leaves three possibilities, and those can also be ruled out individually. Interestingly, the smallest OMSOS that does NOT have a square common sum happens to be unique in another sense, namely, all the entries are squared primes:

$$11^2 \quad 23^2 \quad 71^2$$

$$61^2 \quad 41^2 \quad 17^2$$

$$43^2 \quad 59^2 \quad 19^2$$

The common sum of the rows and columns is 5691 = 3*7*271. Obviously we can permute the rows and columns of an OMSOS without affecting the sums, but since 3*7*271 is not 3 times a square, we know this can't be permuted into a fully magic square. Still, this is an interesting square in its own right. The next two "all-prime" OMSOS's (after the one noted above) are based on the matricies

13	67	149		17	71	149
89	127	53		89	137	29
137	79	43		139	61	67

It's also interesting that the next two "exceptional" OMSOS's (meaning those whose common sum is NOT a square) also have common sums of the form $3*7*p$ where p is a prime congruent to 1 (mod 6).

Even though the OMSOS's with square common sums are immediately excluded from being completely magic, they are interesting in their own right, and it's worthwhile to consider why the condition of equal sums for the row and columns predisposes the common sum to be a square (when the elements themselves are squares). First, notice that they seem to occur in infinite families, and it's not too hard to figure out parametric representations for some of them. For example, there's an infinite family containing "$(1)^2$":

$(1)^2$	$(8+10k)^2$	$(4+16k+10k^2)^2$
$(4+8k+6k^2)^2$	$(4+14k+8k^2)^2$	$(7+8k)^2$
$(8+14k+8k^2)^2$	$(1+8k+6k^2)^2$	$(4+6k)^2$

with the common sum $(9+16k+10k^2)^2$. (Of course there are a few values of k for which the elements of this array are not distinct, so I exclude those from the set of orthomagic squares. Note also that k can be positive or negative, because all the results are squared anyway.) Similarly an infinite family containing $(2)^2$ is given by

$(2)^2$	$(14+10k)^2$	$(5+14k+5k^2)^2$
$(11+10k+3k^2)^2$	$(2+10k+4k^2)^2$	$(10+8k)^2$
$(10+10k+4k^2)^2$	$(5+10k+3k^2)^2$	$(10+6k)^2$

with the common sum $(15+14k+5k^2)^2$. We can give a similar infinite 1-parameter family containing $(n)^2$ for any given n, so there ought to be a 2-parameter representation covering all of these. Ideally we'd like to find a complete characterization of all OMSOS's, or at least the possible common sums, to see if complete magicality can be ruled out.

Orthomagic Square of Squares

The class of orthomagic squares whose "common sum" is a square is closely related to quaternions, spatial rotation matricies, and representations of numbers as sums of FOUR squares. This is an observation that essentially goes back to Euler (see Dickson's History). Specifically, for any numbers a,b,c,d we can construct a 3x3 matrix

$$\begin{matrix} a^2 + b^2 - c^2 - d^2 & 2(bc - ad) & 2(ac + bd) \\ 2(ad + bc) & a^2 - b^2 + c^2 - d^2 & 2(cd - ab) \\ 2(bd - ac) & 2(ab + cd) & a^2 - b^2 - c^2 + d^2 \end{matrix}$$

Each row and column, regarded as a 3D vector, has the magnitude

$$L = a^2 + b^2 + c^2 + d^2$$

so obviously if we construct a 3x3 square whose components are the squares of the components of the above matrix, it will be an "orthomagic square of squares" with the common sum L^2. This accounts for the frequent occurrance of OMSOS's with a square common sum. For example, with a=1, b=2, c=4, d=6 we have the basic matrix

$$\begin{matrix} -47 & 4 & 32 \\ 28 & -23 & 44 \\ 16 & 52 & 17 \end{matrix}$$

and if we square each number this is the smallest orthomagic square of squares, with the common sum $3249 = 57^2$. The Determinant is 57^3.

Note that the three row vectors constitute an orthogonal triad, as do the three column vectors, and if we normalize each term by dividing it by the magnitude L=57 the above matrix is the rotation operator representing the space rotation relating the row triad to the column triad.

Obviously the product of two such base squares gives another base square.

For example, we have the product

$$\begin{vmatrix} -47 & 4 & 32 \\ 28 & -23 & 44 \\ 16 & 52 & 17 \end{vmatrix} \begin{vmatrix} -51 & 18 & 46 \\ 42 & -19 & 54 \\ 26 & 66 & 3 \end{vmatrix} = \begin{vmatrix} 3397 & 1190 & -1850 \\ -1250 & 3845 & 178 \\ 1810 & 422 & 3595 \end{vmatrix}$$

The second factor on the left side is given by setting a=1, b=3, c=5, and d=6, it's determinant is 71^3, and the common sum of squares of its rows and columns is 71^2. The matrix product is also a base square, i.e., the squares of its elements form an orthogonal orthomagic square of squares, with the common sum (57*71)^2 and the base square has determinant (57*71)^3.

It is produced by setting a=-61, b=-1, c=15, d=10, which can be inferred from the four-square multiplication formula

(A^2 + B^2 + C^2 + D^2)(a^2 + b^2 + c^2 + d^2) = w^2 + x^2 + y^2 + z^2

where

w = Aa + Bb + Cc + Dd
x = Ab - Ba + Cd - Dc
y = Ac - Bd - Ca + Db
z = Ad + Bc - Cb - Da

The above is an interesting example of how, when trying to work in three parameters (or dimensions), it often seems that we're led to a much more natural formulation by going to four. I had been looking at sums of THREE squares, noting that the the most general solution (according to Dickson) of the equation X^2 + Y^2 + Z^2 = N^2 is of the form

X = 2(a^2 + b^2 - c^2)
Y = 2(a^2 - b^2 + c^2) + 2a(b-3c)
Z = (-a^2 + b^2 + c^2) + 2b(2a-3c)
N = 3(a^2 + b^2 + c^2) - 2c(2a+ b)

and in retrospect it's clear that there's a 4-parameter family hidden behind this, trying to get out. Indeed, it was derived from Euler's four-square product formula, supressing one of the parameters.

Orthomagic Square of Squares

Incidentally, the quaternionic formulas noted above are very similar to the generalized Heron's formula, relating the volume of a "perfect tetrahedron" in terms of the areas of its faces, as discussed in Heron's Formula For Tetrahedrons.

Anyway, the OMSOS's with square common sum seem to be well covered by the above parameterization, although I'm not quite sure it necessarily includes ALL OMSOS's with square sum. Another interesting question is whether the OMSOS's that *don't* have a square sum are also given by the 4-parameter matrix above, with non-integer values of a,b,c,d.

In other words, are there any sets of numbers a,b,c,d such that the nine elements of the basic matrix are integers but the magnitude of the row and column vectors is not an integer?

There are 91 primitive orthomagic squares of squares with common sums less than 30000. Of those, 56 have a square common sum, whereas the remaining 35 do not. Of the 35 non-square cases, none of them is of the form $3k^2$, so they clearly can't give a complete magic square of squares.

25

Magic Square of Squares

It's an open question whether there exists a 3x3 magic square comprised entirely of square integers. Before considering the possibility of such a square, it's worthwhile to review some basic facts about arbitrary 3x3 magic squares, defined as an array of numbers

$$\begin{array}{ccc} A & B & C \\ D & E & F \\ G & H & I \end{array}$$

such that each row, column, and main diagonal has the same sum.

Letting S denote the common sum, we have the eight defining conditions

$$A+B+C=S \quad D+E+F=S \quad G+H+I=S \quad A+E+I=S$$
$$A+D+G=S \quad B+E+H=S \quad C+F+I=S \quad C+E+G=S$$

Adding up the four conditions involving E, and re-arranging terms, gives

$$(A+B+C) + (D+E+F) + (G+H+I) + 3E = 4S$$

Since each quantity in parentheses equals S, this shows that $3E = S$.

Substituting for S into the four basic equations involving E, we have

Magic Square of Squares

$$A+I = B+H = C+G = D+F = 2E$$

If we subtract 2E from both sides of A+I=2E (for example) we get (A-E) + (I-E) = 0, which shows that A-E = E-I, and likewise for the other cases, proving that each row, column, and diagonal containing E is an arithmetic progression. If we define n = A-E and m = C-E, the magic square contains the terms

$$\begin{array}{ccc} E+n & & E+m \\ & E & \\ E-m & & E-n \end{array}$$

From this the remaining terms follow, due to the requirement that each row and column sum to 3E, so the entire magic square can be expressed in terms of the three numbers E,m,n as shown below:

$$\begin{array}{ccc} E+n & E-n-m & E+m \\ E-n+m & E & E+n-m \\ E-m & E+n+m & E-n \end{array} \quad (1)$$

Now, suppose we imagine a magic square comprised entirely of square integers. This implies the existence of integers E,m,n such that the magic square given by (1) has the values

$$\begin{array}{ccc} a^2 & b^2 & c^2 \\ d^2 & e^2 & f^2 \\ g^2 & h^2 & i^2 \end{array}$$

Equating these to their equivalents in (1) gives the relations

$$\begin{array}{lll} a^2 = E-n & b^2 = E-n-m & c^2 = E+m \\ d^2 = E-n+m & e^2 = E & f^2 = E+n-m \\ g^2 = E-m & h^2 = E+n+m & i^2 = E-n \end{array} \quad (2)$$

Summing the pairs symmetrical about E, we have

$$a^2 + i^2 = b^2 + h^2 = c^2 + g^2 = f^2 + d^2 = 2e^2 \quad (3)$$

Furthermore, if we multiply the pairs symmetrical about E we find that e^4 is expressible as a sum of two squares in the following four ways

$$(ai)^2 + n^2 = e^4$$
$$(bh)^2 + (n+m)^2 = e^4$$
$$(cg)^2 + m^2 = e^4$$
$$(df)^2 + (n-m)^2 = e^4$$

This shows that e^2 is the hypotenuse of four different Pythagorean triples, just as is $2e^2$ (shown by the relations (3)). In addition, if we square both sides of the previous relation $a^2 + i^2 = 2e^2$ (for example) we get

$$a^4 + 2(ai)^2 + i^4 = 4e^4$$

so we can solve this for $(ai)^2$ and substitute into the above expression for e^4, and likewise for the other three expressions, to give

$$a^4 + i^4 = 2[e^4 + n^2]$$
$$b^4 + h^4 = 2[e^4 + (n+m)^2]$$
$$c^4 + g^4 = 2[e^4 + m^2]$$
$$d^4 + f^4 = 2[e^4 + (n-m)^2]$$

This is a very strong set of conditions. It requires a fourth power (e^4) which, if increased by any one of four distinct squares, equals half a sum of two fourth powers. Furthermore, the roots of the four distinct squares must be of the form n, m, n+m, and n-m.

We can also infer more Pythagorean relations by multiplying together other pairs of relations from (2). For example, if we multiply b^2 by f^2 and re-arrange the terms, and likewise for the other pairs, we get

$$(bf)^2 + n^2 = (E-m)^2$$
$$(fh)^2 + m^2 = (E+n)^2$$
$$(hd)^2 + n^2 = (E+m)^2$$
$$(db)^2 + m^2 = (E-n)^2$$

Squaring both sides of the relation $b^2 + f^2 = 2(E-m)$ and solving for $(bf)^2$, we can substitute into the first of these

Magic Square of Squares

equations, and likewise for the other pairs, to give

$$b^4 + f^4 = 2[(e^2 - m)^2 + n^2]$$
$$f^4 + h^4 = 2[(e^2 + n)^2 + m^2]$$
$$h^4 + d^4 = 2[(e^2 + m)^2 + n^2]$$
$$d^4 + b^4 = 2[(e^2 - n)^2 + m^2]$$

Recall that we previously derived the conditions

$$b^4 + h^4 = 2[(e^2)^2 + (n+m)^2]$$
$$d^4 + f^4 = 2[(e^2)^2 + (n-m)^2]$$

so we have found that a necessary condition for a magic square of squares is that there must be four 4th powers b^4, d^4, f^4, h^4 whose sums in pairs each equal twice a sum of squares. This also leads to relations such as

$$f^4 - d^4 = 2(n-m)e^2 \qquad h^4 - b^4 = 2(n+m)e^2$$

However, all these relations just follow algebraically from the more basic conditions, so to prove that no magic square of squares is possible we should probably focus on the basic conditions. Recall that the relations (2) imply

$$\begin{array}{ll} a^2 + i^2 = 2e^2 & b^2 + f^2 = 2g^2 \\ b^2 + h^2 = 2e^2 \qquad f^2 + h^2 = 2a^2 & \qquad (4) \\ c^2 + g^2 = 2e^2 & h^2 + d^2 = 2c^2 \\ d^2 + f^2 = 2e^2 & d^2 + b^2 = 2i^2 \end{array}$$

The four equations on the left show that $2e^2$ must be a sum of two squares in four different ways, so e itself is a sum of squares in (at least) two different ways. The first non-trivial occurrence is $e = 65 = 5*13$, because this is the smallest number that is the product of distinct primes congruent to 1 (mod 4). Each of the factors, in turn, is necessarily a sum of two squares, i.e.,

$$5 = 1^2 + 2^2 \qquad 13 = 2^2 + 3^2$$

The two expressions for 65 arise from the two ways of multiplying these according to the ancient formula

$$(X^2 + Y^2)(A^2 + B^2) = (XB +- YA)^2 + (XA -+ YB)^2 \qquad (5)$$

Taking X=1, Y=2, A=2, B=3 we have

$$e = 65 = (5)(13) = 1^2 + 8^2 = 4^2 + 7^2$$

The four ways of expressing e^2 as a sum of two squares arise from the four ways of multiplying these two forms using equation (5) (and applying the factor $2 = 1^2 + 1^2$, which doesn't change the number of expressions). Thus we have

$$2(65)^2 = 2(1^2 + 8^2)(1^2 + 8^2) = 47^2 + 79^2$$

$$= 2(1^2 + 8^2)(4^2 + 7^2) = 23^2 + 89^2 = 35^2 + 85^2$$

$$= 2(4^2 + 7^2)(4^2 + 7^2) = 13^2 + 91^2$$

We can assemble these squares into a square of squares such as

13^2	23^2	47^2
35^2	65^2	85^2
79^2	89^2	91^2

which has the required equal sums on all four of the lines through the center square (the two main diagonals, the center row, and the center column). Notice, however, that the outer rows and columns do not give the required sum. In general, we can show that

PROPOSITION 1: Any square whose elements satisfy the central sums and whose central number is expressible as a sum of two squares in no more than four distinct ways will NOT give the required sums for the outer rows and columns.

This proposition doesn't completely rule out the existence of a magic square of squares, because it doesn't cover possible squares whose central numbers are expressible as a sum of two squares in MORE than four distinct ways. However, it does

Magic Square of Squares

show that the root of the central number of any magic square of squares would have to be the product of more than two distinct primes of the form 4k+1.

To prove this proposition, note that the above conditions can be expressed parametrically in terms of X,Y,A,B (which had the values 1,2,2,3 in the above example). If we define

$$u = YA - XB \qquad r = YB - XA$$
$$v = YB + XA \qquad s = YA + XB$$

then we have squares such as

$(2uv-uu+vv)^2 \qquad (ur+vs+vr-us)^2 \qquad (2rs+rr-ss)^2$

$(us+vr-vs+ur)^2 \qquad (uu+vv)^2 \qquad (us+vr+vs-ur)^2$

$(2rs-rr+ss)^2 \qquad (ur+vs-vr+us)^2 \qquad (2uv+uu-vv)^2$

which automatically has equal sums on the diagonals and the central row and column. The common sum along each of these lines is

$$S = 3(u^2 + v^2) = 3(r^2 + s^2) = 3[(A^2 + B^2)(X^2 + Y^2)]^2$$

Any magic square of squares whose central element is a sum of two squares in no more than four ways must consist of some arrangement of the four pairs of opposite terms shown in the square above, but they need not be arranged as shown above. It's convenient to refer to each of the outer elements by the negative term it contains. For example, we will let US denote the quantity $(ur+vs+vr-us)^2$, and we will let RR denote $(2rs-rr+ss)^2$, and so on. Thus we have the four pairs of opposite squares (UU,VV), (RR,SS), (US,VR), and (UR,VS).

We note that the system is symmetrical under exchange of (u,v) with (r,s), and we consider first the possibility that UU is in the same row or column with two of the "crossed" elements. In order for the square to be magic the sum of these three quantities must equal the common sum S given above. However, if we examine all six of these case we find that

$UU+VR+VS-S = 8XY(A+B)(A-B)(v+u)(v-u)$
$UU+VR+US-S = 4vu(v+u)(v-u)$
$UU+VR+UR-S = 32ABXYuv$
$UU+VS+US-S = 8uv(A+B)(A-B)(X+Y)(X-Y)$
$UU+VS+UR-S = 4vu(v+u)(v-u)$
$UU+US+UR-S = -8AB(X+Y)(X-Y)(v+u)(v-u)$

In order for the terms of the overall square to be distinct, the values of u,v,r,s must all be non-zero, and the quantities $u^2 - v^2$ and $r^2 - s^2$ must also be non-zero. Furthermore, each of the quantities A,B,X,Y must be non-zero, as must be the quantities $A^2 - B^2$ and $X^2 - Y^2$. These last two condition are due to the fact that if X=Y, for example, we have u=X(A-B), v=X(A+B), r=X(B-A), and s=X(A+B), which implies v=s and u=-r. Inserting these expressions for v and u into ur+vs+us-vr gives $-r^2 + s^2 - 2rs$, so we have VR=SS.

It follows from all these conditions that none of the six quantities shown above can equal zero, so no arrangement with any of those combinations of terms in any row or column can be a magic square of squares. Of course, this also rules out any squares with rows or columns containing SS with any two of the cross terms, since those appear on the opposite side of the above squares.

It remains to consider the possibility of a magic square with each of the outer rows and columns containing two "pure" terms (i.e., UU, VV, RR, SS) and one cross term. This implies that the pure terms must be at the four corners. Therefore, one of the rows or columns will be [UU * RR] where "*" denotes one of the four cross terms. Two of these cases can be ruled out immediately, because we have

$UU+VS+RR-S = 4XY(X+Y)(X-Y)(3A^2 - B^2)(A^2 - 3B^2)$
$UU+UR+RR-S = 4XY(X+Y)(X-Y)(A^2 + 4AB + B^2)(A^2 - 4AB + B^2)$

The first is impossible because a non-zero square cannot be 3 times a square, and the second is impossible because it implies

$$A = +-2B[2+-sqrt(3)]$$

The only two remaining cases are [UU VR RR] and [UU US

Magic Square of Squares 181

RR], each of which corresponds to either of the two squares (because we can transpose the placement of the remaining pair of terms). To prove that these are impossible, we can make use of the right-hand relations in (4), which state that the double of each corner equals the sum of the middle terms of the two opposite sides. This leads us to examine the sums shown below for the indicated square

```
UU VR RR
VS    UR    VR+UR-2SS = 4(s+r)(s-r)[3AB(Y^2 - X^2) + XY(B^2 - A^2)]
SS US VV    UR+US-2UU = 4(v+u)(v-u)[3XY(B^2 - A^2) - AB(Y^2 - X^2)]
```

In order for this to be a magic square, both of the indicated sums must vanish, which (for distinct terms) requires the trailing factors to vanish. But these are of the form 3m+n=0 and 3n-m=0 and therefore m=n=0, which is to say $AB(Y^2 - X^2) = XY(B^2 - A^2) = 0$. This is impossible for distinct terms.

The three remaining cases can be ruled out similarly, based on the sums shown below:

```
UU VR RR
UR    VS    VR+VS-2SS = 4rs[3(A^2 - B^2)(Y^2 - X^2) + 4 ABXY]
SS US VV    VS+US-2UU = 4uv[(A^2 - B^2)(Y^2 - X^2) - 12 ABXY]

UU US RR
VS    UR    US+UR-2SS = 4rs[(A^2 - B^2)(Y^2 - X^2) + 12 ABXY]
SS VR VV    UR+VR-2UU = 4uv[3(A^2 - B^2)(Y^2 - X^2) - 4 ABXY]

UU US RR
UR    VS    US+VS-2SS = 4(s+r)(s-r)[AB(Y^2 - X^2) + 3XY(B^2 - A^2)]
SS VR VV    VS+VR-2UU = 4(v+u)(v-u)[3AB(X^2 - Y^2) + XY(B^2 - A^2)]
```

This completes the proof. So, what has been shown here is that if the central number $2e^2$ is expressible as a sum of two squares in just four distinct ways (which is the minimum number necessary to satisfy the diagonal and center row and column conditions), then it's impossible to satisfy the outer row and column conditions. Of course, every array will be based on just four partitions of $2e^2$ into two squares, but hose partitions won't all necessarily come from the same factorization of e, so those cases aren't covered by the above argument.

26

Anti-Carmichael Pairs

Let the integers a,b be called an "anti-Carmichael pair" if a divides b(b-1) and if b divides a(a-1). This definition was suggested by George Baloglou, who gave the example a=63,b=217. Here's one way of characterizing all such pairs: We want positive integers A,B such that A divides B(B-1), and B divides A(A-1), which implies integers M,N such that

$$A(A-1) = MB \qquad B(B-1) = NA$$

Let c be the greatest common divisor of A and B, so we have A=ac, B=bc where gcd(a,b)=1. The above equations become

$$a(ac-1) = Mb \qquad b(bc-1) = Na$$

Since a and b are coprime we know that a divides M, and b divides N, so there are integers n,m such that M=ma and N=nb. Thus the above equations reduce to ac-1 = mb and bc-1 = na, which can be written as

$$ac - bm = 1 \qquad bc - an = 1$$

Since a and b are coprime we are guaranteed infinitely many solutions of each of these equations. The solutions of the left hand equation will be of the form $c = c\_0 + bk$ and of the right hand will be $c = c\_0 + aj$, and so the solutions to both

Anti-Carmichael Pairs

equations are of the form $c = c_0 + (ab)t$ for any integer t. The value of c_0 can be found via the Euclidean Algorithm applied to either of the above equations. Incidentally, also we have the linear equation

$$a^2 n - b^2 m = (b-a)$$

Anyway, this shows that $(A,B) = (ac,bc)$ is an anti-Carmichael pair if and only if

$$A = a(c_0 + ab\ t) \qquad B = b(c_0 + ab\ t)$$

for any non-negative integer t, where c_0 is the least positive solution
of $ac-bm=1$. Thus for ANY pair of coprime integers (a,b) we can construct the unique infinite family of anti-Carmichael pairs.

For example, let's take a=9 and b=31. The least positive solution of $9c-31m=1$ is with c=7, so it follows that any pair of integers of the form (9(7+279t), 31(7+279t)) is anti-Carmichael for any non-negative integer t. With t=0 this gives George's (63,217), whereas with t=1 it gives (2574,8866), and so on.

27

Coherent Arrays of Squares

Given a set of n real numbers $a_1, a_2, ..., a_n$ we can define an n-dimensional array **A** of size m^n by

$$A[i_1, i_2, ..., i_n] = C + \sum_{j=1}^{n} i_j a_j$$

$$i_1, i_2, ..., i_n = 0, 1, ... m-1$$

where C is an arbitrary constant and m is a positive integer. We'll refer to an array of this kind as *coherent*. If we regard the components of such array as the values of a continuous function of the indices, then the defining characteristic is that the partial derivative with respect to each index is a constant, i.e.,

$$\frac{\partial A}{\partial i_j} = a_j$$

For example, given two real numbers a_1, a_2 and a contant C, and taking m=3, we have a 3x3 array of the form

C	$C + a_2$	$C + 2a_2$
$C + a_1$	$C + a_1 + a_2$	$C + a_1 + 2a_2$
$C + 2a_1$	$C + 2a_1 + a_2$	$C + 2a_1 + 2a_2$

Coherent Arrays of Squares

A coherent array with n=1 is simply an arithmetic progression of length m, whereas a coherent array with m=1 is just a single constant for any n.

It's interesting to consider, for various dimensions n and sizes m, whether there exist coherent arrays with those parameters whose components are all square integers. We know there are sets of three squares in arithmetic progression, but there are <u>no</u> four squares in arithmetic progression. Therefore, we can say that there exist coherent arrays of squares of dimension n=1 and of size m=1,2,3, but not of size m=4 or greater.

If we go on to consider arrays of dimension n=2, it's easy to see that there exist coherent arrays of squares of size m=1 and 2. For example, we have the array

$$1^2 \quad 4^2$$
$$7^2 \quad 8^2$$

However, it's less clear whether there exists a 2-dimensional coherent array of squares of size m=3. The existence of such an array is equivalent to the existence of a 3´3 magic square of squares.

For dimension n=3 the first non-trivial case to consider is arrays of size m=2. It turns out that there do exist coherent arrays of squares with n=3,m=2, such as

$$10^2 \quad 50^2 \qquad 94^2 \quad 106^2$$
$$67^2 \quad 83^2 \qquad 115^2 \quad 125^2$$

This array is generated by the values $a_1 = 2400$, $a_2 = 4389$, $a_3 = 8736$, and $C = 100$. The eight components can be pictured as lying at the eight vertices of a cube. Obviously given any such array we can create another by simply multiplying each component by a fixed square. All the primitive solutions with greatest partial derivative less than 200000' .

The next case to consider is arrays of dimension n=4 and size m=2. To seach for these, we can simply take each coherent array of squares with n=3,m=2 and check to see if there is any integer a_4 which, when added to each component, gives a square. None of the n=3,m=2 solutions listed in the Attachment gives a solution of this kind. So, we have another open question. It may be that no coherent array of squares of dimension n=4 and size m greater than 1 exists.

The table below summarizes what is known about coherent arrays of squares of various dimensions and sizes. The symbol "Y" signifies that arrays exist, "N" signifies that thay do not exist, and "(?)" means none are known but the existence has not been disproven.

| | | m | | | |
n	1	2	3	4	5
1	Y	Y	Y	N	N
2	Y	Y	(?)		
3	Y	Y	(?)		
4	Y	(?)			
5	Y				

Obviously if no arrays exist for a given n,m, then none exist for any pair of parameters greater than or equal to n and m respectively.

28

Mock-Rational Numbers

For any positive integer n let f(n) denote the integer given by the recurrence

$$f(n) = 10 f(n-1) + n$$

with the initial value f(0) = 0. Thus we have f(1)=1, f(2)=12, f(3)=123, and so on. The square roots of f(n) for odd integers n give a persistent pattern, appearng to be rational for periods, and then disintegrating into irrationality. This is illustrated by the first 500 digits of sqrt[f(49)] below.

sqrt[f(49)] =

1111111111111111111111111.1111111111111111111111
0860
5555555555555555555555555555555555555555555555
2730541
6666666666666666666666666666666666666666
0296260347
22222222222222222222222222222222222222
0426563940928819
4444444444444444444444444444444
38775512504011171874
999999999999999999999999999
8082 9687711486305338541

6666666666666666666666
59871857386214406386555598958
33333333333333333333
0843460407627608206940277099609374
99999999999999
06422275875559830666394303215874565970
222222222
18634920167911808330818440.....

Notice how the sequence of digits consists of strings of repeated digits alternating with "scrabled" strings. The repeating strings become progressively smaller and the scrabled strings become larger until eventually the repeating strings disappear. However, by increasing n we can forstall the disappearance of the repeating strings as long as we like.

The repeating digits are always

1, 5, 6, 2, 4, 9, 6, 3, 9, 2, ...

To determine this sequence of digits directly, recall that for any positive integer k the sum of $(n^t)(x^n)$ can be expressed in terms of a polynomial of degree k whose coefficients are a row of the Eulerian numbers. In the present case we just have t=1 so it we immediately have

$$f(n) = (10^{(n+1)} - 9n - 10)/81$$

Setting n=2k-1 (because n must be odd) and factoring gives

$$f(2k-1) = \frac{10^{(2k)}}{81} \left(1 - \frac{18k+1}{10^{(2k)}} \right)$$

The square root of this number is

$$\sqrt{f(2k-1)} = \frac{10^k}{9} \left(1 - \frac{18k+1}{10^{(2k)}} \right)^{1/2}$$

Expanding the square root gives

$$\sqrt{f(2k-1)} = \frac{10^k}{9}\left[1 - \frac{1}{2}\left(\frac{18k+1}{10^{(2k)}}\right) - \frac{1}{8}\left(\frac{18k+1}{10^{(2k)}}\right)^2 - \ldots\right]$$

The repeated digits can be inferred from the k=0 case, which implies that each term contributes a negative digit equal to the ultimate repeating digit of $(c_t)/9$, where c_t is the coefficient of $((18k+1)/(10^{(2k)}))^t$ in the binomial expansion.

Beginning with the first repeating digit $a_0 = d_0 = 1$, we have $c_1 = 1/2$ and so the 1st order term contributes $(-1/2)/9 = -.05555555$.

Accordingly we set $a_1 = 5$ and subtract this from the initial repeating digit $d_0 = 1$. Thereafter we set d_j to the least positive (non-zero) residue (mod 9) of $d_{(j-1)} - a_j$. For example, since $a_1 = 5$ and $d_0 = 1$ we compute $d_0 - a_1 = 1 - 5 = -4$, and so we set $d_1 = 5$.

In this way we can compute the successive digits that will appear in strings just by examining the ultimate repeating digits of the binomial expansion coefficients divided by 9. This gives the values

i	c_i	a_i	d_i
0	1	1	1
1	1/2	5	5
2	1/8	8	6
3	1/16	4	2
4	5/128	7	4
5	7/256	4	9
6	21/1024	3	6
7	33/2048	3	3
8	$429/2^{15}$	3	9
9	$715/2^{16}$	7	2

etc.

To more efficiently generate the terms of this sequence, notice that binomial expansion coefficients are the partial products (up to sign) of

$$\frac{1}{2}, \frac{1}{4}, \frac{3}{6}, \frac{5}{8}, \frac{7}{10}, \frac{9}{12}, \frac{11}{14}, \frac{13}{16}, \frac{15}{18}, \frac{17}{20}, \frac{19}{22}, \frac{21}{24}, \frac{23}{26}, \ldots$$

Since every odd factor appears in the numerator no later than it appears in the denominator, the denominators of the partial sums are necessarily pure powers of 2. Also, since the numerators are all odd, the denominator has exactly the powers of 2 cumulatively of the first k integers, given by the sequence

$$0,1,3,4,7,8,10,11,15,16,\ldots$$

whose terms can be generated by $E(n) = n + E([n/2])$. Since $1/2 = 5 \bmod 9$, the effect of the denominators on the partial products mod 9 is to multiply by $5^{\wedge}E(n) \pmod 9$. The cycle of powers of 5 (mod 9) is $1,5,7,8,4,2,1,\ldots$, which a period of 6. Therefore, if we let p[j] with $j=0,1,..,5$ denote the array [1,5,7,8,4,2], the nth denominator contributes a factor of p[E(n) mod 6]. This already gives a non-periodic sequence, and explains why every expansion of one of these coefficients divided by 9 ends in a single repeating digit (which would not be the case if the coefficients could have factors other than 2 in the denominator).

To properly account for the numerators as well as the denominators, we need to keep track of the cumulative powers of 3 in the fraction (taking numerator powers as positive and denominator powers as negative), since we are working modulo 9. Given the value of a_j representing the jth partial product (mod 9) and the cumulative powers of 3 in this product, and given the (j+1)th fractional factor N/D, we first factor out the powers of 3 from N and D, and increment/decrement the cumulative total accordingly, to give the new total p_3(j) and the fraction n/d where n and d represent the 3-free parts of N and D respectively. Then we can reduce these (mod 9) and take the reciprocal of d (mod 9) to form the multiplier.

The products of these 3-free fractions gives a sequence f_j of positive non-zero integers (mod 9). Then the value of a_j is defined as

Mock-Rational Numbers

$$a_j = \begin{cases} f_j & \text{if } P_j = 0 \\ 3 f_j \bmod 9 & \text{if } P_j = 1 \\ 0 & \text{if } P_j > 1 \end{cases}$$

It's not hard to see that the sequence a_j contains infinitely many zeros, because $a_{\{3^k - 1\}} = 0$ for all k. Also, since $a_{\{3^k\}} = k-1$, it's clear that a_j attains arbitrarily large ranges consisting of nothing but zeros. This gives another demonstration of the fact that a_j (and therefore d_j) is not periodic. Here's a simple BASIC program for computing arbitrarily many values of a_j:

```
DEFDBL A-Z
DIM recip(8)
FOR i = 0 TO 8: READ recip(i): NEXT i
DATA 0,1,5,0,7,2,0,4,8
f = 5: sum3 = 0
FOR nn = 1 TO 5000 STEP 2
n = nn
d = nn + 3
DO WHILE ((n MOD 3) = 0)
  n = n \ 3
  sum3 = sum3 + 1
LOOP
DO WHILE ((d MOD 3) = 0)
  d = d \ 3
  sum3 = sum3 - 1
LOOP
n = n MOD 9
d = d MOD 9
f = (f * n * recip(d)) MOD 9
IF sum3 = 0 THEN fout = f
IF sum3 = 1 THEN fout = (f * 3) MOD 9
IF sum3 > 1 THEN fout = 0
PRINT nn, f, fout
INPUT tttt
NEXT nn
```

Here are the first 502 values of a_j:

1584743337175250003336660007178586662824740000000000003336660006663330000000000007178586665251410006663330002821413337175250000000000000000000000000000000000000003336660006663330000000000006663330003336660000000000000000000000000000000000000007178586665251410006663330005257173334748580000000000006663330003336660000000000002821413334748580003336660007178586662824740000000000000000000000000000000000000000000000000000000000000000000000000000000000000000000000000000000000000000000000000003336660006663330...

Here are the corresponding 502 values of d_j, which are the digits that repeat for progresively smaller bursts in our original "mock-rational numbers":

15624963921375999963936999921348936978624999999999999963936999936963999999999999213489369426519999369639999786519639213759999999999999999999999999999999999996393659993696399999999999936963999𝜍539369999999999999999999999999999999999213489369426519999369639999426879639573489999949999999369639999639369999999999997865196395734899996393699 99

2134893697862499999999999999
9999999999999999999999999999
9999999999999999999999999999
9999999999999999999999999999
9999999999999999999999999999
6393699993696399...

Interestingly, if we take these as the digits of a single number, it gives another example of what might be called a "mock-rational number", because it has arbitrarily long ranges of pure 9's, but infinitely many non-9 digits.

In some respects, the most interesting sequence is f_j, which is the same as a_j above, except that we simply discard the effects of the powers of 3 in the binomial expansion coefficients. Here are the first 502 values of the f_j sequence, arranged in blocks of 27:

158474141717525525717282474
717858525282474474858141858
717858858858717858141858474
717858282525141717858141282
282141474717525525141858141
858474474474858717282717858
717858282525141474282717525
858474717141282141282717282
141525525525141858141858474
717858282525141141525474858
525717141474858282474525474
858474474474858474525474282
282141717474858282141858717
717858525282474474858141858
141525525525141282717282141
858474141717525232141858717
474282858525141282474525474
282141141141282717282717858
7178582825251411...

This highlights the quasi-periodicity of the digits when taken in blocks of 3^k. I wonder if there are plane-tiling patterns of simple shapes that correspond to this quasi-periodic 3^k pattern. Each of the digits 1,2,4,5,7,8, appears (asymptotically) an equal number of times in this pattern. One way of visualizing this peculiar sequence is to assign each of the six possible digits to one of six directions in the plane, and then plot the visited locations reached by a "pseudo-random" walk produced by this sequence. The first 50 million digits lead to the locus of points shwon below in blue. The red circle was the starting point, and the red dot was the ending point (i.e., the 50 millionth step).

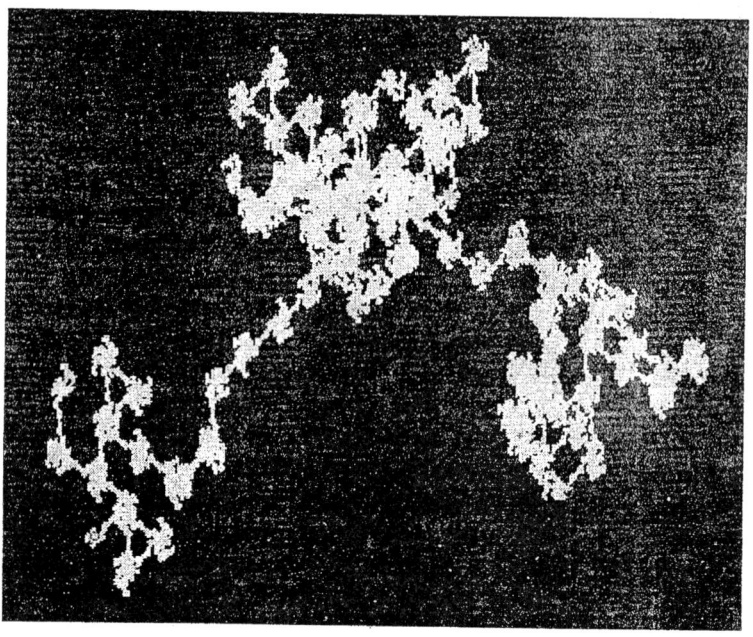

It's fascinating to watch this "pseudo-random" walk progress, noting that it carefully retraces its steps at various stages, which explains why it has branches that appear to lead nowhere. The figure below shows the orbit extended to cover the first 100 million digits, with the color changing every 5 million digits.

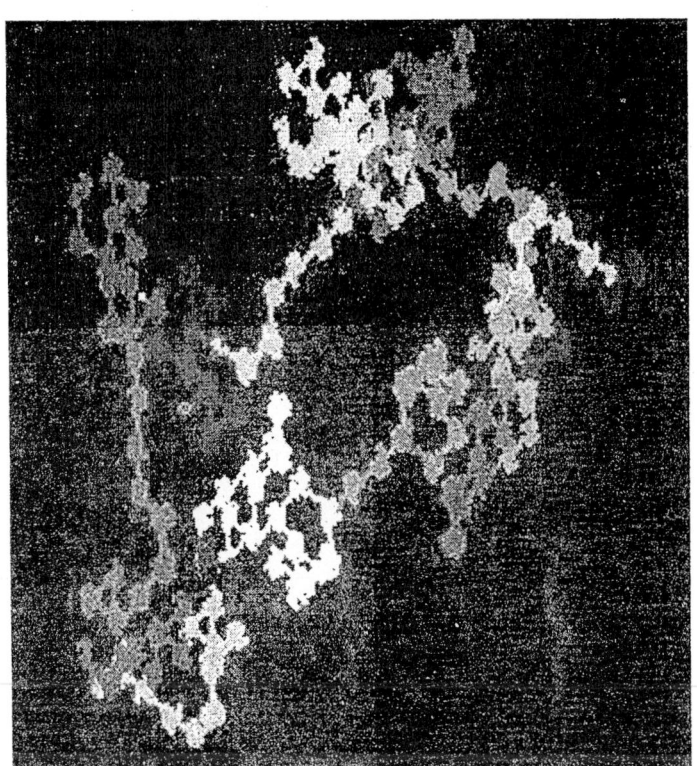

29

Integer Sequences Related To PI

For any given integers s[0] and s[1] we can construct the infinite sequence s[n], n=0,1,2,... generated by the recurrence

$$s[k+2] = (2k+3) s[k+1] + (k+1)^2 s[k] \qquad (1)$$

For example, with the initial values s[0]=0, s[1]=1 we produce the sequence

0, 1, 3, 19, 160, 1744, 23184, 364176, 6598656, 135484416, ...

On the other hand, if we take the initial values s[0]=1, s[1]=0 we produce the sequence

1, 0, 1, 5, 44, 476, 6336, 99504, 1803024, 37019664, ...

For convenience, let [m,n;k] denote s[k] based on the initial values s[0]=m and s[1]=n. It turns out that the two sequences listed above can serve as the "basis" for all sequences of this form, according to the relation

$$[m,n;k] = (m)[1,0;k] + (n)[0,1;k]$$

Anyway, the ratios of corresponding terms in these sequences approach some interesting values. For example, we can show that for any integers m,n (positive or negative)

Integer Sequences Related To PI

$$\lim_{k \to \inf} \frac{[m,n;k]}{[1,1;k]} = \frac{n-m}{4} PI + m \qquad (2)$$

Simply dividing two general expressions of this form, we get the more general relation for any w,x,y,z

$$\lim_{k \to \inf} \frac{[x,y;k]}{[w,z;k]} = \frac{(y-x)PI + 4x}{(z-w)PI + 4w} \qquad (3)$$

To illustrate, note that the sequence [1,1;k] has the values

1, 1, 4, 24, 204, 2220, 29520, 463680, 8401680, 172504080, ...

and with m=-2, n=+1 the sequence [m,n;k] has the values

-2, 1, 1, 9, 72, 792, 10512, 165168, 2992608, 61445088, ...

The ratio of the 9th terms of these sequences is

$$\frac{61445088}{172504080} = 0.356194984... \sim \frac{3}{4} PI - 2$$

If we add 2 to this ratio and multiply by 4/3 we get 3.141593...

Another interesting observation concerns the growth rate of the individual sequences. It appears that the difference in consecutive ratios of consecutive values approaches 1 + sqrt(2). In other words

$$\lim_{k \to \inf} \frac{[m,n;k+2]}{[m,n;k+1]} - \frac{[m,n;k+1]}{[m,n;k]} = 1 + \sqrt{2} \qquad (4)$$

I suppose these results apply to any real values of m,n, not just integers, so equation (3) gives the limit of ratios of terms of any sequence [x,y;k] to terms of the sequence [w,z;k]. I'd be interested if anyone could supply an explicit solution of the recurrence (1), which is a simple homogeneous linear 2nd-order recurrence, but with non-constant coefficients.

By the way, since (3) shows that all the sequences are easily related, it's enough to consider just the ratio of two sequences. Define the integer sequences D[k] and N[k] such that both satisfy the same recurrence relation, namely

$$N[k+2] = (2k+3) N[k+1] + (k+1)^2 N[k]$$

and

$$D[k+2] = (2k+3) D[k+1] + (k+1)^2 D[k]$$

with the initial values

$$N[0] = 0 \qquad D[0] = 1$$
$$N[1] = 1 \qquad D[1] = 1$$

Thus the values of N and D are

$$N = \{0, 1, 3, 19, 160, 1744, 23184, 364176, 6598656,...\}$$
$$D = \{1, 1, 4, 24, 204, 2220, 29520, 463680, 8401680,...\}$$

It can be shown that the ratio N[k]/D[k] equals the kth convergent of the continued fraction for the arctangent of 1:

$$PI/4 = \cfrac{1}{1 + \cfrac{1}{3 + \cfrac{4}{5 + \cfrac{9}{7 + \cfrac{16}{9 + \cfrac{25}{\text{etc.}}}}}}}$$

Incidentally, this continued fraction can also be inferred from the fact that

Integer Sequences Related To PI

$$PI = -2i \ln\left(\frac{i-1}{i+1}\right)$$

Anyway, the continued fraction can be converted into the following infinite series

$$\frac{PI}{2} = 1 + \frac{3}{4} - \frac{5}{24} + \frac{7}{204} - \frac{9}{1408} + \frac{11}{10455} - \frac{13}{71484} + \frac{15}{478429} - \frac{17}{3148884} + \ldots$$

where the kth term c_k is just the difference between the (k-1)th and the (k+1)th convergent of the continued fraction. In other words

$$c_k = \frac{N[k+1]}{D[k+1]} - \frac{N[k-1]}{D[k-1]}$$

As a result, the partial sums of the series are just the sums of two consecutive convergents of the continued fraction. Anyway, it's nice that the sequence of numerators is just the odd numbers. It would be even nicer to have an explicit expression for the denominators. In that direction, notice that if we let g denote the greatest common divisor of

$$D[k-1]\, D[k+1] \quad \text{and} \quad D[k+1]\, N[k-1] - D[k-1]\, N[k+1]$$

it follows that

$$D[k+1]\, N[k-1] - D[k-1]\, N[k+1] = (2k+1)\, g$$
$$\text{and}$$
$$D[k-1]\, D[k+1] = d_k\, g$$

where d_k is the denominator of the kth term in the above series for PI/2. Now it appears that

$$D[k+1]\, N[k-1] - D[k-1]\, N[k+1] = (2k+1)\, [(k-1)!]^{\wedge}2 \qquad (5)$$

so $g = [(k+1)!]^{\wedge}2$, which implies that

$$d\_k = \frac{D[k-1] \; D[k+1]}{[(k-1)!]^2}$$

Anyway, (5) is sort of a convolution of the terms of the two sequences, which is in some way a more natural definition of their relation than the original linear 2nd order recurrence. Actually (5) is just one of a set of formulas relating the "cross-products" of these two series. In general we have

$$N[k] \; D[k+j] - N[k+j] \; D[k] = [k!]^2 \; P(k)$$

where $P(k)$ is a polynomial in k of degree j-1. The first few such polynomials are listed below

j	P(k)
1	1
2	3 + 2 k
3	19 + 20 k + 5 k^2
4	160 + 214 k + 90 k^2 + 12 k^3
5	1744 + 2718 k + 1497 k^2 + 348 k^3 + 29 k^4
6	23184 + 40336 k + 26453 k^2 + 8236 k^3 + 1225 k^4 + 70 k^5

The constant coefficients are just the values of the N[] sequence, whereas the leading coefficients 1, 2, 5, 12, 29, 70, ... are the "Pell numbers" that satisfy the recurrence s[n]=2s[n-1]+s[n-2].

Also, notice that the sums of the coefficients for the jth polynomial are 1, 5, 44, 476, 6336, 99504, ... which are the values of the other "basis sequence" mentioned previously. In other words, these values satsify the same recurrence as do N[] and D[], but with the initial values 1,0,..., which is the sequence [1,0;k].

By the way, A.P.Magnus comments that

$$D[k+1] = 2 \; (k+1)! \; P\_k(i) \; /i^k \; , \quad k=1,2,...$$

where P_k is the k-th degree Legendre polynomial, and N[k] is an associated Legendre polynomial.

One last observation. Using the original sequence notation, we have the extremely interesting result:

$$\left| \frac{[0,1;k]}{[-1,0;k]}, 1, k \right| = 0$$

30

Series Within Parallel Resistance Networks

If electrical resistors r1 and r2 are connected in series they give a total resistance of r1 + r2. Likewise connecting r3 and r4 in series gives a total resistance of r3 + r4. Putting these series combination in parallel gives a combined resistance of

$$rx = \frac{(r1 + r2)(r3 + r4)}{r1 + r2 + r3 + r4}$$

Quentin Grady notes that if we connect the node between r1 & r2 to the node between r3 & r4 the circuit is transformed to a parallel within series connection, and the resistance becomes

$$ry = \frac{r1\ r3}{r1 + r3} + \frac{r2\ r4}{r2 + r4}$$

He asks for positive integer solutions of this equation, i.e., integers values of r1, r2, r3, r4, rx, and ry, "preferably under 100", that satisfy this equation.

There are a very large number of solutions to this. Even if we require r1 through r4 to be distinct, and add the restriction

Series Within Parallel Resistance Networks   203

that the values of (r1*r3)/(r1 + r3) and (r2*r4)/(r2+r4) must both be integers, there are still an over-abundance of solutions. Most of them have r1*r4 = r2*r3, which implies that rx=ry. In other words, the overall resistance is unchanged by inserting the jumper. An example is [r1,r2,r3,r4] = [3,9,6,18]. You can easily define infinite families of such solutions. For example, with

$$r1 = 8j+2 \quad r2 = 24j+8k+10 \quad r3 = 24j+6 \quad r4 = 72j+24k+30$$

we have rx = ry = 24j + 6k + 9. On the other hand, there are also many solutions with rx > ry. An example is [5,99,95,9].

There are so many solutions to these equations that I think it's more interesting to restrict the problem a little more by requiring not only that Rx and Ry are integers but also the components of Ry, i.e., R1R3/(R1+R3) and R2R4/(R2+R4). In other words, we seek four integers a,b,c,d such that each of the quantities

$$M = \frac{ac}{a+c} \qquad N = \frac{bd}{b+d} \qquad K = \frac{(a+b)(c+d)}{a+b+c+d}$$

is an integer. The expressions for M and N imply

$$(a-M)(c-M) = M^2 \qquad (b-N)(d-N) = N^2$$

This means there are integers W,X,Y,Z such that WX=M^2, YZ=N^2, and

$$a = M+W \qquad b = N+X \qquad c = M+Y \qquad d = N+Z$$

Thus for any integers M,N we need only select factorizations of M^2 and N^2 into two factors to generate values of a,b,c,d that satisfy the equations for M and N. Now we just need to ensure that the equation for K is also satisfied.

The equation for K implies the existence of integers m,n such that mn=K^2 and

$$(a+b) = K + m \qquad (c+d) = K + n$$

Thus we have

$$(a+b) - (c+d) = m - n \qquad (1)$$
$$(a+b) + (c+d) - 2K = m + n \qquad (2)$$
$$K^2 = mn \qquad (3)$$

Letting Q denote $a+b+c+d-2K$, equations (2) and (3) imply that m and n are the roots of

$$m^2 - Qm + K^2 = 0$$

so we have

$$m = \frac{Q + \sqrt{Q^2 - 4K^2}}{2} \quad n = \frac{Q - \sqrt{Q^2 - 4K^2}}{2}$$

To make these integers it's clear that the quantity in the radical must be a square, so there is an integer f such that

$$f^2 + (2K)^2 = Q^2$$

By Pythagorean triples this implies the existence of integers q,r,s such that

$$f^2 = q(r^2 - s^2) \qquad K = qrs \qquad Q = q(r^2 + s^2)$$

and so we can express m and n as

$$m = \frac{q(r^2 + s^2) + q(r^2 - s^2)}{2} = qr^2$$

$$n = \frac{q(r^2 + s^2) - q(r^2 - s^2)}{2} = qs^2$$

Substituting these expressions for m, n, and K back into equations (1) and (2) gives

$$q(r^2 - s^2) = (a+b) - (c+d) \qquad (4)$$

Series Within Parallel Resistance Networks

$$q(r + s)^2 = (a+b) + (c+d) \qquad (5)$$

Now recall that we have

$$a = M+W \qquad b = N+Y \qquad c = M+X \qquad d = N+Z$$

where $WX=M^2$ and $YZ=N^2$. It follows that there must be integers e,f,w,x,y,z such that

$$W = ew^2 \qquad X = ex^2 \qquad Y = fy^2 \qquad Z = fz^2$$

and so $M = ewx$ and $N = fyz$. Substituting these expressions into equations (4) and (5) gives

$$q(r^2 - s^2) = (ewx+ew^2 + fyz+fy^2) - (ewx+ex^2 + fyz+fz^2)$$

$$q(r + s)^2 = (ewx+ew^2 + fyz+fy^2) + (ewx+ex^2 + fyz+fz^2)$$

Cancelling terms, these equations reduce to

$$q(r+s)(r-s) = e(w^2 - x^2) + f(y^2 - z^2)$$

$$q(r+s)(r+s) = e(w + x)^2 + f(y + z)^2$$

Multiplying the first by $(w+x)$ and the second by $(w-x)$, and then subtracting one form the other gives

$$q(r+s)[(w+x)(r-s) - (w-x)(r+s)] = f(y+z)[(w+x)(y-z) - (w-x)(y+z)]$$

which reduces to

$$q(r+s)(xr-ws) = f(y+z)(xy-wz)$$

Similarly we can eliminate f from the two preceeding equations to give the relation

$$q(r+s)(zr-ys) = e(w+x)(wz-xy)$$

This shows that there are two general classes of solutions to the overall problem. The degenerate case is when

$$xr-ws = xy-wz = zr-ys = 0$$

which is true iff $(r/s) = (w/x) = (y/z)$. In this case we can set $w=jy$ and $x=jz$ for any rational j, and we have the parametric set of solutions

$$\begin{aligned} a &= ewx+ew^2 = ey(y+z)j^2 & M &= y\,z\,e\,j^2 \\ b &= fyz+fy^2 = fy(y+z) & N &= y\,z\,f \\ c &= ewx+ex^2 = ez(y+z)j^2 & K &= y\,z\,(ej^2+f) \\ d &= fyz+fz^2 = fz(y+z) \end{aligned}$$

On the other hand, if the quantities $xr-ws$, $xy-wz$, and $zr-ys$ are not zero, we can still have solutions by solving the equations

$$q(r+s)(xr-ws) = f(y+z)(xy-wz)$$

$$q(r+s)(zr-ys) = e(w+x)(wz-xy)$$

for e, f, and q. Clearly the first equation is satisfied if we set $q = (y+z)(xy-wz)$ and $f = (r+s)(xr-ws)$. Substituting this value of q into the second equation gives

$$(y+z)(r+s)(zr-ys)(xy-wz) = e(w+x)(wz-xy)$$

or

$$e = \frac{(y+z)(r+s)(ys-zr)}{(w+x)}$$

To ensure that e is an integer we can multiply e, f, and q by $(w+x)$ to give the solution

$$\begin{aligned} e &= (r+s)(y+z)(ys-zr) \\ f &= (w+x)(r+s)(xr-ws) \\ q &= (y+z)(w+x)(xy-wz) \end{aligned}$$

Any common factors can be divided out of these three numbers. Then the final form of the solution (in the nondegenerate case, and without reducing to lowest terms) is

$$a = w(ys-zr)J \qquad M = (r+s)(y+z)(ys-zr)wx$$
$$b = y(xr-ws)J \qquad N = (r+s)(x+w)(xr-ws)yz$$
$$c = x(ys-zr)J \qquad K = (x+w)(y+z)(xy-wz)rs$$
$$d = z(xr-ws)J$$

where $J = (x+w)(y+z)(r+s)$. By the way, notice that

$$Rx - Ry = K - N - M = (ys-zr)(xr-ws)(xy-wz)$$

so this is the amount by which the two circuits differ. The smallest solutions are those for which each of the three values $ys-zr$, $xr-ws$, and $xy-wz$ have their minimum possible value, which is $+1$, so most of the small solutions give $Rx - Ry = +1$.

There's also a connection with what are called "Farey sequences". Remember that in order to give all positive resistances we require each of the quantities $ys-zr$, $xr-ws$, and $xy-wz$ to be positive, which is the case if and only if

$$(y/z) > (r/s) > (w/x)$$

Now, observe that the four basic resistors a,b,c,d will be proportional to w,y,x,z if we can ensure that the "commutators" are all $+1$. All we really need to do is take any four integers w,x,y,z such that $xy-wz=1$ and then set $r=w+y$ and $s=x+z$. This automatically ensures not only that r/s is between y/z and w/x, but that the quantities $xr-ws$ and $ys-zr$ are both equal to 1. This automatically gives a solution of the simple form

$$a = wJ \qquad M = (r+s)(y+z)wx$$
$$b = yJ \qquad N = (r+s)(x+w)yz$$
$$c = xJ \qquad K = (x+w)(y+z)rs$$
$$d = zJ$$

where $J = (x+w)(y+z)(r+s)$ and the difference $Rx-Ry$ equals 1. Of course, the difference between Rx and Ry is also equal to

$$\frac{(ad-bc)^2}{(a+c)(b+d)(a+b+c+d)}$$

which shows (again) why the difference cannot be negative.

31

Pythagorean Graphs

Is it possible to fill an infinite square array with distinct integers such that the sum of the squares of any two adjacent numbers is a square? To illustrate, the following is a 4x4 array with the desired property

1836	105	252	735
1248	100	240	700
936	75	180	525
273	560	1344	3920

Every pair of neighboring numbers (horizonally or vertically) constitutes the legs of a Pythagorean triple. The hypotenuses of these triples are as shown below

```
        1839      273      777
   2220      145      348      1015
        1252     260      740
   1560      125      300      875
        939      195      555
   975      565     1356      3955
        625      1456     4144
```

Pythagorean Graphs

The key to this problem is to recognize that the row and column conditions are independent. Thus, for any positive integer N we can easily construct a square array of size (N+1)x(N+1) consisting of the values

$$A[m,n] = (12/5)^m \; (24/7)^n \; 35^N$$

$$= 2^{(2m+3n)} \; 3^{(m+n)} \; 5^{(N-m)} \; 7^{(N-n)}$$

for m,n = 0,1,2,..,N. These values are all integers and no two of them have the same number of factors of 5 and 7, so they are all distinct.

Although this gives a recipe for arbitrarily large arrays, it doesn't give an *infinite* array. R. Mentock provided the following nice construction of an infinite array.

For m=3M+a, n=3N+b, define F(m,n) by

$$F(m,n) = 6^M * 132^N * f(a,b)$$

where

f(0,0) = 6 * 7	f(0,1) = 6 * 24	f(0,2) = 6 * 143
f(1,0) = 8 * 7	f(1,1) = 8 * 24	f(1,2) = 8 * 143
f(2,0) = 15* 7	f(2,1) = 15* 24	f(2,2) = 15* 143

Another, shorter, solution: For m=2M+a, n=2N+b, define F(m,n) by

$$F(m,n) = 10^M * 76^N * f(a,b)$$

where

f(0,0) = 7 * 17	f(0,1) = 7 * 144
f(1,0) = 24 * 17	f(1,1) = 24 * 144

Both of the above "infinite" solutions actually fill only one quadrant of the plane. To completely "tile" the plane with Pythaogrean legs.

Each of the two solutions are actually combinations of two sequences, each defined along the axes. The other points are just products of these, i.e., G(m,n) = G(m) * G(n). Since we have four of them, we start at zero and assign values to the four semi-axes. G(0,0) is the LCM of all G(0), and that number is propogated through. Then G(m,n) = G(m) * G(n) like above, and there we go.

A slightly different approach was suggested by Dave Radcliffe, who wrote

Given F(m,n) = 10^M * 76^N * f(a,b) where

f(0,0) = 7 * 17   f(0,1) = 7 * 144
f(1,0) = 24 * 17   f(1,1) = 24 * 144

For m,n >= 0 define

G(m,n)   = F(m,n) * 11 * 13
G(-m-1,n)   = F(m,n) * 11 * 84
G(-m-1,-n-1) = F(m,n) * 60 * 84
G(m,-n-1)   = F(m,n) * 60 * 13

Since no F(m,n) is divisible by 11 or 13, G is also one-to-one.

By the way, it's also interesting to see which polyhedra can have distinct numbers at their vertices such that the sum of the squares of any two adjacent numbers is a square. Considering just the Platonic solids, I think there is no solution for the tetrahedron, octahedron, or icosahedron. Of course there are solutions for the cube, such as

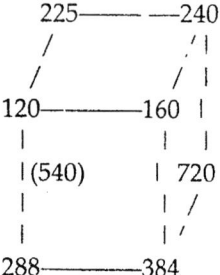

Pythagorean Graphs

Here the values increase by factors of 4/3, 12/5, and 15/8 in the three principle directions. (Presumably it's possible to fill an infinite 3D orthogonal lattice with distinct numbers in this way.)

This leaves only the dodecahedron. It turns out that it is possible to populate the vertices of a dodecahedron with distinct numbers such that the sum of the squares of each pair of adjacent numbers is a square. Here's an example of a "Pythagorean dodecahedron":

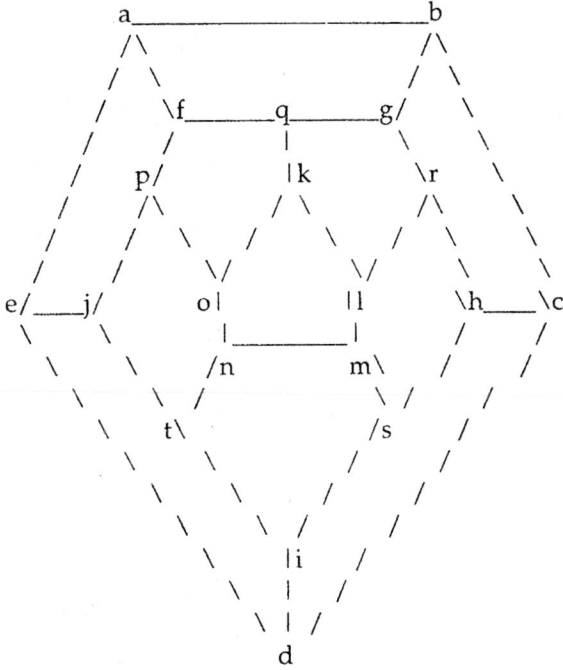

a = 155040   f = 649800   k = 802560   p = 2006400
b = 290700   g = 1438965   l = 601920   q = 180576
c = 897600   h = 3762000   m = 451440   r = 689700
d = 1683000   i = 2613600   n = 240768   s = 846450
e = 177650   j = 1504800   o = 1070080   t = 275880

Another related question, raised by Dave Radcliffe, is whether it it possible to connect any two positive integers (greater than 4) by a series of Pythagorean legs. (As Dave put it, "consider a graph G with a vertex for each integer n > [4], and an edge joining m and n iff $m^2 + n^2$ is a square. Is this graph connected?")

Notice that the numbers 3,4 are a separate graph and don't connect to anything else. But for n > 4 it certainly appears to be true that the graph is connected.

If you look at this as a special case of the more general "connection criterion" $m^2 + n^2 = k*$square (where k is a squarefree integer expressible as a sum of two squares), it seems the case k=1 is unique in giving a connected graph for all the integers (above some number).

In contrast, consider the graph based on $m^2 + n^2 = 2*$square. In this case the naturals split into mutually disjoint graphs, each of which is a multiple of the basic connected graph consisting of the numbers

1, 7, 17, 23, 31, 41, 47, 49, 71, 73, 79, 89, 97, 103, 113, 119, ...

These are exactly the numbers whose prime factors are all congruent to +1 or -1 modulo 8. In other words, they are the primes p such that 2 is a square (mod p).

The graph based on $m^2 + n^2 = 5*$square splits into mutually disjoint graphs, each of which is a multiple of

1, 2, 4, 8, 11, 16, 19, 22, 29, 31, 32, 38, 41, 44, 58, 59, 61, 62,...

These appear to be exactly the numbers expressible as products of the primes 2, 11, 19, 31, 41, 59, 61, 71, 79, 89, 101, ... which I think are the primes p such that 5 is a square (mod p).

The graph based on $m^2 + n^2 = 10*$square splits into disjoint graphs that are each multiples of

1, 3, 9, 13, 27, 31, 37, 39, 41, 43, 53, 67, 71, 79, 81, 83, ...

Again it appears there are all the products of a particular

set of primes, in this case the primes 3, 13, 31, 37, 41, etc., and these look like the primes modulo which 10 is a square.

So, we might be tempted to conclude that the graph of the naturals with the connection criterion $m^2 + n^2 = k*\text{square}$ splits into disjoint graphs, each of which is a multiple of a graph consisting of the numbers expressible as a product of primes p not dividing k and such that k is a quadratic residue mod p. (This explains why the case k=1 gives just a single totally connected graph.) However, the case k=13 seems to behave differently. It's basic connected set consists of

1, 8, 12, 18, 27, 34, 51, 53, 79, 86, 92, 96, 103, 116, 122, ...

The set of primes for this case should be 2,3,13,17,23,29,43,etc. but there seems to be something else going on here.

32

On General Palindromic Numbers

Patrick De Geest asked for tetrahedral numbers, i.e., numbers of the form $(b*(b+1)*(b+2))/6$, that are also palindromic (digits read the same forwards and backwards). He noted that $b=336$ gives the decimal palindrome 6378736. Of course, if we don't restrict ourselves to decimal representations we can find many such numbers. For example, the base-9 representation of the tetrahedral number $k(k+1)(k+2)/6$ is palindromic for the following values of k (shown in decimal):

k	$k(k+1)(k+2)/6$ (in base 9)
1	1
2	4
3	11
4	22
12	444
13	555
120	488884
588	71066017
1093	505555505
88573	505055555550505

Needless to say, the "tetrahedral" numbers are just the binomial coefficients $C(k,3)$. The base-3 representations of $C(k,3)$ are pal .dromic for the following values f k

k	C(k,3) (in base 3)
1	1
2	11
3	101
4	202
6	2002
12	111111
14	202202
39	112121211
120	112222222211
392	201000222000102
496	1102111111112011

Similarly we have

$C(1105,3)$ = 1534114351 in base 8

$C(521,3)$ = 1122123212211 in base 4

$C(1166,3)$ = 6364334636 in base 7

$C(82332,3)$ = 12 5 9 13 18 18 13 9 5 12 in base 27

There is also a tetrahedral number C(k,3) whose base-B representation is palindromic where both k and B are perfect squares (k=441, B=16).

Of course, every positive integer has a palindromic representation in SOME base. If we let f(n) denote the smallest base relative to which in is palindromic, then clearly f(n) is no greater than n-1, because every number n has the palindromic form '11' in the base (n-1).

Interestingly, for almost all integers n the value of f(n) is actually quite small, as can be gathered from the table below:

n	f(n)	n	f(n)	n	f(n)	n	f(n)	n	f(n)
1	2	21	2	41	5	61	6	81	8
2	3	22	10	42	4	62	5	82	3
3	2	23	3	43	6	63	2	83	5
4	3	24	r	44	10	64	7	84	11

n	f(n)	n	f(n)	n	f(n)	n	f(n)	n	f(n)
5	2	25	4	45	2	65	2	85	2
6	5	26	3	46	4	66	10	86	6
7	2	27	2	47	46	67	5	87	28
8	3	28	3	48	7	68	3	88	5
9	2	29	4	49	6	69	22	89	8
10	3	30	9	50	7	70	9	90	14
11	10	31	2	51	2	71	7	91	3
12	5	32	7	52	3	72	5	92	6
13	3	33	2	53	52	73	2	93	2
14	6	34	4	54	8	74	6	94	46
15	2	35	6	55	4	75	14	95	18
16	3	36	5	56	3	76	18	96	11
17	2	37	6	57	5	77	10	97	8
18	5	38	4	58	28	78	5	98	5
19	18	39	12	59	4	79	78	99	2
20	3	40	3	60	9	80	3	100	3

The numbers for which $f(n) = n-1$ are

3, 4, 6, 11, 19, 47, 53, 103, 137, 139, 149, 163, 167, 179, 223, 263, 269, 283, 293, 311, 317, 347, 359, 367, 389, 439, 491, 563, 569, 593, 607, 659, 739, 827, 853, 877 977, 983,...

These might be called "the non-palindromic numbers" because they are the only numbers n that are NOT palindromic in ANY base from 2 to n-2.

Proposition: If $f(n) = n-1$ for $n > 6$ then n is a prime.

Proof: If $n=ab$ with $a < (b-1)$ and $b > 2$ we have $n=a(b-1)+a$, so n has the palindromic form 'aa' when expressed in the base $(b-1)$.

This covers all composites greater than 6 except for squared primes, but for squares we have $a^2 = (a-1)^2 + 2(a-1) + 1$, so every square $n > 4$ has the palindromic form 121 in the base $\sqrt{n}-1$. Done.

The ratio of non-palindromic primes to all primes drops as the range increases. For example, of the 1229 primes less than 10000 we find that 218 are non-palindromic, which is about 1 out of every 5.6. Of the 9295 primes less than 100000 we find only 1199 are non-palindromic, which is only about 1 out of every 7.75.

At the opposite extreme, there are primes that are already palindromic in the base 2, although these are even more scarce

On General Palindromic Numbers

than non-palindromic primes. The first few are

3, 5, 7, 17, 31, 73, 107, 127, 257, 313, 443, 1193, 1453, ...

Returning to the function f(n) for general values of n (not restricted to primes), it's clear that f(n), n=1,2,3,... fluctuates between 2 and n-1, but I think the average value of f(n) for all n from 1 to infinity is polynomial in n. Specifically it appears that

$$f(n)\_average = n^{(1/c)}$$

where c is a constant approximately equal to 2.68. In other words,
I conjecture that

$$\lim_{n \to \inf} \frac{1}{n} \sum_{k=1}^{n} \frac{\ln(k)}{\ln(f(k))} = c$$

but I don't know how to prove that this limit actually exists.

Anyway, it's also interesting to observe the values of f(n) when n is a prime power. For example, powers of 2 have the following values of f(n):

n	f(n)	representation of n in the base f(n)					
2	3	2					
4	3	1	1				
8	3	2	2				
16	3	1	2	1			
32	7	4	4				
64	7	1	2	1			
128	7	2	4	2			
256	15	1	2	1			
512	7	1	3	3	1		
1024	7	2	6	6	2		
2048	31	2	4	2			
4096	7	1	4	6	4	1	
8192	15	2	6	6	2		
16384	15	4	12	12	4		

In general it appears that the min-base palindromic representation of 2^m is a multiple of a binomial expansion, i.e.,

$$2^m = 2^r [(2^s - 1) + 1]^t$$

Obviously we have $m = st + r$. The values of r,s,t for the first several m are tabulated below:

m	r s t	m	r s t	m	r s t
1	1 2 0	11	1 5 2	21	1 5 4
2	0 2 1	12	0 3 4	22	2 5 4
3	1 2 1	13	1 4 3	23	2 7 3
4	0 2 2	14	2 4 3	24	0 6 4
5	2 3 1	15	0 5 3	25	0 5 5
6	0 3 2	16	0 4 4	26	1 5 5
7	1 3 2	17	1 4 4	27	3 6 4
8	0 4 2	18	3 5 3	28	0 7 4
9	0 3 3	19	1 6 3	29	1 7 4
10	1 3 3	20	0 4 5	30	0 5 6

In each case the triple (r,s,t) is such that s is minimized under the constraint that

$$2^s - 1 > \begin{cases} (2^r) * C(t,t/2) & \text{if } t \text{ is even} \\ (2^r) * C(t-1,(t-1)/2) & \text{if } t \text{ is odd} \end{cases}$$

The min-base palindromic representation of other prime powers also tend to be multiples of binomial expansions, but not always. For example, the min-base representation of 3^7 is [3 19 3] in the base 24. Also, the min-base representation of 7^6 is [1 2 3 2 1] in the base 18. This raises the question of whether the min base representation for powers of 2 is always of the binomial form. Can anyone supply a proof or counter-example?

33

Minimizing the Denominators of Unit Fraction Expansions

For any proper fraction n/d let f(n,d) denote the least integer k such that n/d can be expressed as a sum of distinct unit fractions with denominators less than or equal to kd. For example, we have f(31/311) = 10, because 31/311 can be expressed as a sum of distinct unit fractions with denominators no greater than 3110, but not with any smaller max denominator. I'm seeking a reference, proof, or counter-example for the conjecture that f(n,d) < 2 ln(d) + 1.

The proposition would certainly be false if the +1 was omitted, because we have f(7,19) = 6, whereas 2*ln(19) equals only 5.89.

On the other hand, with the +1 term the proposition is true for all d < 4831. For example, the formula correctly asserts that the fraction 43/4021 can be expressed as a sum of distinct unit fractions with greatest denominator 68357.

It certainly appears that highest values of max{f(n,d)} are nearly linear as a function of log(d), and on probabilistic grounds it's easy to see that the relation should be of this form. I'd like to know if the asymptotic slope is really 2, or if something steeper is required.

According to the abstract, this paper proves that, in my notation,

$$f(n,d) \le (\log d)^{3/2} + \epsilon$$

where epsilon goes to zero as d goes to infinity.

However, I've found a counter-example to my conjecture. The unit fraction expansion of 5/6947 with the smallest maximum denominator is

$$\frac{1}{6947}\left(\frac{1}{2}+\frac{1}{3}+\frac{1}{4}+\frac{1}{5}+\frac{1}{7}+\frac{1}{8}+\frac{1}{13}+\frac{1}{14}+\frac{1}{19}\right)+\frac{1}{3640}+\frac{1}{5187}$$

so we have f(5,6947) = 19, whereas my conjectured upper bound is 2 log(6947) + 1 = 18.692.. If we define D(d) as max{f(n,d) | 0 < n < d} then the smallest values of d for each D(d) < 20 are listed below:

D(d)	smallest d	D(d)	smallest d
1	2	11	271
2	3	12	419
3	5	13	521
4	11	14	751
5	13	15	1423
6	19	16	2273
7	41	17	4021
8	37	18	5659
9	89	19	6947
10	179		

It's interesting to consider why it's impossible to express every simple fraction of the form n/19 (for example) as a sum of distinct unit fractions with max denominator of 5*19 or less. The reason can be explained in terms of the first five reciprocals (mod 19). These five numbers are

$$\begin{aligned} 1/1 &= 1 \\ 1/2 &= 10 \\ 1/3 &= 13 \quad (\text{mod } 19) \\ 1/4 &= 5 \\ 1/5 &= 4 \end{aligned}$$

In order to express every fraction n/19, n = 1 to 18, as a sum of unit fractions with denominators no larger than 5*19 we

Minimizing the Denominators of Unit Fraction Expansions

would need to be able to express each n from 1 to 18 as a sum (mod 19) of the above five numbers, i.e., 1, 10, 13, 5, 4, without repetition. The possible sums of these numbers (mod 19) are

```
sums of one:     1  10  13   5   4
sums of two:    11  14   6   5   4  15  14  18  17  9
sums of three:   5  16  15   0  18  10   9   8   0  3
sums of four:   13   4   1   9  10
sums of five:   14
```

The numbers 2, 7, and 12 do not appear, so these are the numerators n such that the expansion of n/19 requires a denominator of at least 6*19.

Obviously it's a simple matter to recursively generate the sums of the reciprocals (mod p) of the first k integers, for k=1,2,... until finding a solution. Roughly speaking, if we treat the sums of random samples, we would expect to need $O(\log(p))$ samples to get any particular residue modulo p, so this immediately leads us to suppose that we could expand any simple fraction n/p into a sum of unit fractions with greatest denominator roughly $O(p \log(p))$.

The 1976 paper of Bleicher and Erdos concludes "with a numerical example which illustrates that the algorithms to date [for expanding fractions into sums of unit fractions minimizing the largest denominator] leave something to be desired." They note that the Fibonacci-Sylvester algorithm yields an expansion with huge denominators (although there seems to be a typo in the actual values printed in the paper). In contrast, Erdos's algorithm gives

$$\frac{5}{121} = \frac{1}{48} + \frac{1}{72} + \frac{1}{180} + \frac{1}{1452} + \frac{1}{4354} + \frac{1}{8712} + \frac{1}{87120}$$

They also noted that the continued fraction algorithm gives a shorter expansion with smaller denominators

$$\frac{5}{121} = \frac{1}{25} + \frac{1}{1225} + \frac{1}{3477} + \frac{1}{7081} + \frac{1}{11737}$$

Then they compare this with the algorithm described in their paper, which gives

$$\frac{5}{121} = \frac{1}{30} + \frac{1}{242} + \frac{1}{363} + \frac{1}{1210} + \frac{1}{3630}$$

They then applied "ad hoc" methods to get "a good short expansion"

$$\frac{5}{121} = \frac{1}{42} + \frac{1}{70} + \frac{1}{330} + \frac{1}{5082}$$

and then another expansion "which while longer has denominators considerably smaller than any of the others"

$$\frac{5}{121} = \frac{1}{42} + \frac{1}{70} + \frac{1}{726} + \frac{1}{770} + \frac{1}{1815}$$

It's interesting that, at recently as 1976, people evidently weren't familiar with the simple recursive algorithm for finding the unit fraction expansion with the absolute smallest max denominator. It's easy to see that, since 121 is composite the recursive algorithm can be applied to (1/11)(5/11) to give the almost trivial expansion 5/121 = 1/33 + 1/121 + 1/363. This makes it rather surprising that 5/121 was selected by Erdos and Bleicher to exemplify a "hard" fraction, especially considering that this method was apparently used to construct the tables of unit fractions in the Akhmin Papyrus, c. 500 AD.

By the way, I've done some more checking and found that the lowest order fraction n/d such that D(d)/d = 20 is 1097/14939, which expands to 1/14939 times

$$\left(\frac{1}{1} + \frac{1}{2} + \frac{1}{3} + \frac{1}{4} + \frac{1}{5} + \frac{1}{6} + \frac{1}{7} + \frac{1}{8} + \frac{1}{9} + \frac{1}{10} + \frac{1}{12} + \frac{1}{14} + \frac{1}{16} + \frac{1}{18} + \frac{1}{20} \right)$$

Minimizing the Denominators of Unit Fraction Expansions   223

plus a remainder of $41/560 = 1/14 + 1/560$. Interestingly this is back in agreement with my original conjecture that $D(d)/d < 2 \log(d) + 1$.

Another interesting point is that the only four numerators for which $f(n,14939)$ equals 20 are 1097, 1927, 13235, and 14065. Notice that both pairs differ by 830. Evidently these are the only four numbers that can't be expressed as sums in terms of the reciprocals of the first 20 integers modulo 14393.

We alluded above to a recursive method for expanding a simple fraction n/p into a sum of distinct unit fractions by forming the sums of the reciprocals of the first k integers modulo the prime denominator p.

The first sum that equals the numerator n yields an expansion of the form

$$\frac{n}{p} = \frac{1}{p} \left(\frac{1}{x_1} + \frac{1}{x_2} + .. + \frac{1}{x_j} \right) + \frac{u}{v}$$

where $0 < x_1 < x_2 ... < x_j <= k$ and v is not divisible by p. I said that this method gives the expansion with the least possible max denominator, p^*x_j. Of course, this assumes the remainder u/v can be expanded into a sum of unit fractions with max denominator less than p^*x_j, which is ordinarily the case, because v can only be a product of divisors of the x's, each of which is smaller than roughly $\log(p)$.

However, there are cases in which the remainder of the first solution produced by the recursive formula requires a denominator greater than p^*x_j in its expansion. For example, if we search recursively for an expansion of $3/2221$ the first solution is occurs with k=11, namely

$$\frac{3}{2221} = \frac{1}{2221} \left(1 + \frac{1}{2} + \frac{1}{3} + \frac{1}{4} + \frac{1}{5} + \frac{1}{6} + \frac{1}{7} + \frac{1}{8} + \frac{1}{9} + \frac{1}{11} \right) + \frac{1}{27720}$$

but in this case $p^*x_j = 24431$, which is less than the remainder's denominator of 27720. What we've shown is that the greatest denominator in an expansion of $3/2221$ must be at least 24431, and need be no greater than 27720. To determine if there is an expansion with max denominator between 24431 and 27720 we must proceed to the next solution in the recursion, which occurs at k=13:

$$\frac{3}{2221} = \frac{1}{2221}\left(1 + \frac{1}{2} + \frac{1}{5} + \frac{1}{6} + \frac{1}{7} + \frac{1}{10} + \frac{1}{13}\right) + \frac{1}{2730}$$

This shows that the next smallest possible value of p*xj is 28873, and no later expansion in the recursion sequence can have a lesser max denominator than this. Therefore, the preceeding solution is optimum.

In general, this method of determining the optimum (least max denominator) expansion consists of recursively generating the solutions in increasing order of p*xj until finding one for which the remainder can be expanded with a max denominator less than p*xj.

This almost always occurs on the first solution, but if it doesn't the process continues until such a remainder is found.

Roughly speaking we will always find solutions with k less than $O[\log(p)^{\wedge}(1+\text{delta})]$, and the recurrence involves $2^{\wedge}k$ trials, so the number of trials is very roughly on the order of p. This approach is even more efficient for finding the limiting expansion for ALL the numerators for a given denominator p, becauase this can be computed in essentially the same time required to solve for a single numerator.

Of course the above algorithm applies only to expanding fractions with prime denominators. From the standpoint of determining the upper bound on max denominators this is not a serious limitation because the greatest values of (max denom in expansion)/(denom) are known to occur for prime denominators. However, it could be a limitation in carrying out the above algorithm in cases were the remainder u/v is nontrivial. Fortunately the fact that v is necessarily the product of many distinct small primes implies that it's usually quite easy to find a robust expansion of u/v simply by partitioning u into divisors of v.

34

Perrin's Sequence

Define the sequence of integers by the linear recurrence formula

$$s[n] = s[n-2] + s[n-3]$$

with the initial values $s[0] = 3$, $s[1] = 0$, $s[2] = 2$. This sequence was discussed by Edouard Lucas in 1878, who noted that if p is a prime then p divides $s[p]$. This is an immediate consequence of Fermat's Little Theorem, and as such is a necessary but not sufficient condition for primality. Subsequently (1899) the same sequence was mentioned by R. Perrin (L'Intermediaire Des Mathematiciens). The most extensive (published) treatment of this sequence was given in an excellent paper by Dan Shanks and Bill
    Shanks and Adams referred to this as Perrin's sequence.
    This sequence has many interesting properties, making it, in some ways, more remarkable than the Fibonacci sequence. For example, most people are familiar with the spiral of equilateral squares whose edge lengths correspond to the Fibonacci numbers, but less well-known is the spiral of equilateral triangles shown below described in The Golden Pentagon.

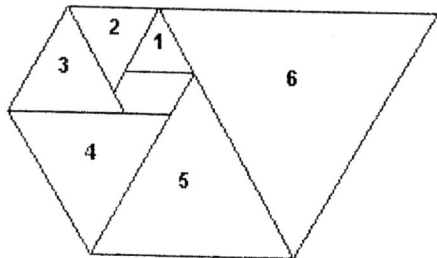

The edge lengths of successive triangles in this spiral satisfy the Perrin recurrence $s[n]=s[n-2]+s[n-3]$ as well as the recurrence $s[n]=s[n-1]+s[n-5]$, as is apparent from the above figure. This can be seen as a consequence of the fact that the characteristic polynomial of Perrin's sequence, $x^3 - x - 1$, is a divisor of the characteristic polynomial of the 5th order sequence, i.e.,

$$x^5 - x^4 - 1 = (x^3 - x - 1)(x^2 - x - 1)$$

Interestingly, the real root r of $x^3 - x - 1$ has the nice expression as a sequence of nested cube roots:

$$r = \sqrt[3]{1 + \sqrt[3]{1 + \sqrt[3]{1 + \sqrt[3]{1 + \ldots}}}}$$

$$= 1.324717957244746\ldots$$

If we define the angle q in terms of this root r as

$$q = \text{invcos}\left(\frac{-1}{2} r^{3/2}\right) = 2.437734932\ldots$$

Perrin's Sequence

then the terms of the Perrin sequence can be expressed in closed form as

$$s[n] = r^n + 2\,\frac{\cos(nq)}{r^{\wedge}(n/2)}$$

In fact, with the appropriate provisos, we can replace n with any complex argument z to define Perrin's function on the complex plane, as shown in the figure below for real components from -5 to 15 and imaginary components from -5 to +5.

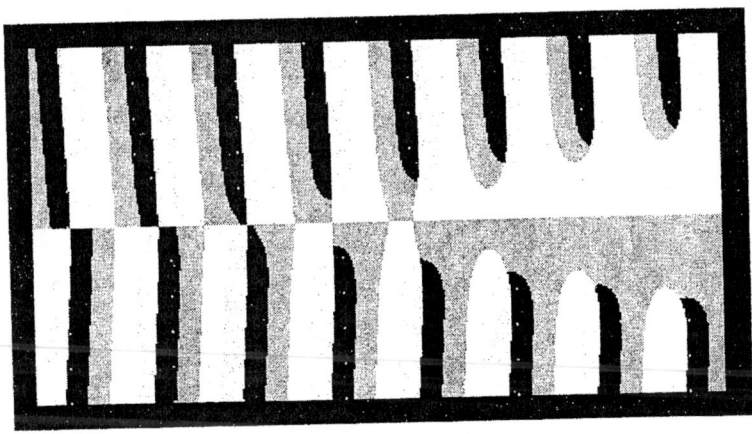

White indicates regions where both the real and imaginary components are positive, black where both are negative, and the two shades of gray where one is positive and the other negative. The zeros of this function are all on the real axis in the left half-plane, whereas there are two sets of complex conjugate zeros on the right half-plane (at points where all four shades meet).

Obviously, for any positive prime p and any integer n we have

$$s[pn] = s[n] \quad (\mathrm{mod}\ p)$$

so in particular we have

$$s[p] = 0$$
$$s[-p] = -1 \quad (\bmod\ p)$$

In addition, for any integers m,n,k we have the relation

$$s[mn+k] = s[m]s[m(n-1)+k] - s[-m]s[m(n-2)+k] + s[m(n-3)+k]$$

Which is really just a special case of a much more general class of relations satisfied by any linear recurring sequence. In this particular case we have relations like

$$s[2n] = s[n]^2 - 2s[-n]$$

$$s[3n] = s[n]^3 - 3s[n]s[-n] + 3$$

and so on, as well as

$$s[2n+1] = s[n]s[n+1] + s[1-n]$$

Using these relations for s[2n] and s[2n+1] gives an efficient means of computing s[k] via the usual binary pattern algorithm on the plus and minus sides of the sequence. This same two-sided approach can be extended to higher order recurrences, but it quickly becomes more practical to use the more general one-sided relations (described in the note mentioned above).

Perrin's sequence also has the interesting property that its terms are cumulative sums of the sequence itself, i.e., we have s[1]=0 and

$$s[n] = \sum_{k=-3}^{n-5} s[k] \quad \text{for } n > 1$$

$$s[n] = \sum_{k=4}^{4-n} s[-k] \quad \text{for } n < 1$$

The terms of Perrin's sequence can also be expressed as a function of binomial coefficients, leading to many interesting results. For example, it's possible to deduce that the summation

$$\sum_{k=0}^{N} \frac{(2N+k)!}{(2N-2k)!\,(1+3k)!}$$

is an integer for N=2755452, and for no smaller N.

As a compositeness test, Perrin's sequence is much stronger than the typical 2nd order Lucas sequence. For example, the smallest symmetric pseudoprime relative to the Fibonacci quadratic $x^2 - x - 1$ is 705, whereas the smallest realtive to Perrin's cubic $x^3 - x - 1$ is

$$27664033 = (3037)(9109)$$

as found by Shanks and Adams (using an HP-41C calculator!). Subsequently Shanks, G. Kurtz, and H. Williams tabulated all the symmetric pseudoprimes relative to Perrin's sequence less than $50(10)^9$. They noted that none of these pseudoprimes had the signature of a prime p such that Perrin's polynomial is irreducible (mod p). As far as I know the question of whether such a pseudoprime exists is still open.

35

Unit Fraction Partitions

Consider the Diophantine equation

$$\sum_{i=0}^{n} \frac{1}{a(i)} = \frac{1}{k}$$

where n, k, and a(i), i=0 to n, are positive integers. Empirically we note that, for fixed n, the number of unordered solution sets for the array a() increases roughly (though not uniformly) as k increases.

For example, with n=2 the number of solution sets for $0 < k < 25$ are

$$3 \quad 10 \quad 21 \quad 28 \quad 36 \quad 57 \quad 42 \quad 70 \quad 79 \quad 96 \quad 62 \quad 160$$
$$59 \quad 136 \quad 196 \quad 128 \quad 73 \quad 211 \quad 80 \quad 292 \quad 245 \quad 157 \quad 93 \quad 366$$

Can the number of solutions for a() be given explicitly in terms of k for any fixed n?

When n equals 2 this asks for the number of partitions of 1/k into three unit fractions. I think a complete answer to this question will be difficult, because it includes as a special case the unproved conjecture (of Erdos and Straus) that 4/k can always be expressed as a sum of three unit fractions. (Clearly any such expression can be divided through by 4 to give 1/k as a sum of three unit fractions.) According to Richard Guy's Unsolved Problems in Number Theory, "there is a go d account

Unit Fraction Partitions

of this problem in Mordell's book, where it is shown that the conjecture is true, except possibly in cases where n is congruent to 1^2, 11^2, 13^2, 17^2, 19^2, or 23^2 mod 840." Also, it's been verified numerically for all $k < 10^8$.

Of course, we can account for many of the partitions of $1/k$ simply in terms of the partitions of $1/j$ where j divides k. For example, of the 366 partitions of $1/24$, only 164 are "primitive" in the sense that the gcd of the three denominators is 1. Letting $U(n)$ denote the number of 3-part unit fraction partitions of n, the non-primitive partitions of $1/24$ can be counted by inclusion-exclusion as follows

$$U(24/2) + U(24/3) - U(24/6) = 160 + 70 - 28 = 202$$

In general, the number of 3-part partitions of $1/k$ can be grouped according to the greatest common divisor (gcd) of their denominators.

The results are summarized in the table below. (For reasons of space I've shown only gcd's up to 21.)

gcd of denominators

k	1	2	3	4	5	6	7	8	9	10	11	12	13	14	15	16	17	18	19	20	21
1	1	1	1																		
2	4	1	2	1	1	1															
3	8	4	1	3	2	1	1	0	1												
4	11	4	2	1	3	2	1	1	1	1	0	1									
5	13	6	6	1	1	3	2	1	0	1	1	0	0	0	1						
6	20	8	4	4	1	1	6	3	2	2	1	1	1	1	0	0	1				
7	12	8	4	4	3	1	1	3	2	0	1	0	0	1	1	0	0	0	0	1	
8	20	11	10	4	3	2	1	1	4	3	2	2	1	1	0	1	1	1	0	1	0...
9	28	10	8	2	6	4	1	1	1	4	3	3	1	1	2	0	0	1	1	0	1...
10	34	13	6	6	4	6	2	1	2	1	5	3	2	2	2	1	1	0	0	1	1...
11	20	10	7	6	2	4	1	0	0	0	1	3	2	1	1	0	1	0	0	0	0...
12	55	20	11	8	9	4	5	4	2	1	2	1	9	6	3	3	2	2	2	2	1...
13	19	11	5	0	3	1	5	3	1	0	1	0	1	3	2	0	0	0	0	1	0...

Letting $G(k,j)$ denote the number of 3-part unit fraction partitions of $1/k$ whose denominators have a gcd of j, we have

$$G(kn, jn) = G(k,j)$$

It follows that, for example, the sum of $G(24,2j)$ for $j=1,2,...$ is equal to the sum of $G(12,j)$ for $j=1,2,...$, which equals $U(12) = 160$.
Similarly the sum of $G(24,3j)$, $j=1,2,...$ is just equal to $U(8) = 28$. Then by inclusion-exclusion we know that the sum of $G(24,j)$ for all j divisible by either 2 or 3 is given by

$$U(12) + U(8) - U(4) = 160 + 70 - 28 = 202$$

By the same approach we see that there are 285 non-primitive partitions of 1/30, given by

$$U3(15) + U3(10) + U3(6) - U3(5) - U3(3) - U3(2) + U3(1)$$
$$= 196 + 96 + 57 - 36 - 21 - 10 + 3 = 285$$

Once the non-primitive partitions have been accounted for, the remaining partitions counted by the terms $G(k,j)$ for $j < 3k$ and j coprime to k. For example, there are 55 primitive partitions of 1/12, and there are 9 primitive partitions of 5/12, and 5 of 7/12, and so on.

It might seem at first that the above table implies no solution of 4/13, because there are no primitive solutions for 1/13 with gcd of 4. However, there are solutions with gcd=8 and gcd=20, so we can multiply these by 4 to express 4/13 as a sum of three unit fractions. However, the table also shows that there are no 3-part partitions of 1/11 with gcd of 8, 16, 24, or 32. It follows that 8/11 cannot be expressed as a sum of three unit fractions.

Another approach is based on the fact that for any positive integer k there exists a least integer $D(k)$ such that the partitions of $1/k$ into three unit fractions correspond to the partitions of $D(k)/k$ into three divisors of $D(k)$. For example, we have $D(1)=12$ because the 3-part unit fraction partitions of 1/1 correspond to the partitions of 12 into three divisors of 12:

$$12 = 6 + 4 + 2 = 6 + 3 + 3 = 4 + 4 + 4$$

Similarly, we have $D(2) = 2520$ because the 3-part unit fraction partitions of 1/2 correspond to the partitions of 1260 into exactly three divisors of 2520:

Unit Fraction Partitions

$$1260 = 840 + 360 + 60$$
$$840 + 315 + 105$$
$$840 + 280 + 140$$
$$840 + 252 + 168$$
$$840 + 210 + 210$$
$$630 + 504 + 126$$
$$630 + 420 + 210$$
$$630 + 315 + 315$$
$$504 + 504 + 252$$
$$420 + 420 + 420$$

The first few values of $D(k)$ are 12, 2520, 65520, 3603600, ... It may be easier to express these in terms of the exponents in their prime factorizations:

k	exponents of $D(k)$
1	21
2	3211
3	421101
4	422111
5	52211110001
6	53221111100001
7	6221101101
8	542111110001100000001
9	532111111011001000001
10	632121111011000001001
11	432111111001001000101
12	6422121111111101010011000000000000001
13	5322111010100010010011
14	7322111111111110001110000010100000000000000001
15	8532211111101111010011100000000000000000000000000001
16	6422112111011000100011100000000001
17	3421111111101010000111000100000000000000000000000000000000000000001
18	74231112111111101100111000000001000000000101
19	7421111111100100001011101000001000000000001
20	74322111111111111111111111011000000000000000000100... (421)

It's clear that $D(k)$ is always divisible by $k+1$. Also, it appears that $D(k)$ is always divisible by $q = k^2 + k + 1$. Moreover, if q is a prime, then it is the largest prime dividing $D(k)$. In any case, the largest prime divisor of $D(k)$ is no greater than q.

Another interesting approach is to place the original equation

$$1/k = 1/a + 1/b + 1/c \tag{1}$$

in "dimensionless" form by simply multiplying through by k:

$$1 = x + y + z \tag{2}$$

where $x=k/a$, $y=k/b$, and $z=k/c$. Equation (1) is just the equation of a diagonal plane in 3D space, and we are looking for rational points on this plane whose numerators divide k. We could try rotating this plane into a 2D space to simplify things (although the rotation might not preserve rationality).

Interestingly, if we make the substitutions

$$X = a/k - 1 \qquad Y = b/k - 1 \qquad Z = c/k - 1$$

then the original equation (1) becomes

$$XYZ - X - Y - Z = 2 \tag{3}$$

This equation seems to arise in many different problems. Here the only effect of k is to determine which of the rational solutions of (3) give integer values of a,b,c according to

$$a = (X+1)k \qquad b = (Y+1)k \qquad c = (Z+1)k$$

In other words, k just "scales up" the rational solutions of (2), resulting in a certain number of them being integers. If k=1 then the only integer solutions (a,b,c) correspond to the positive integer solutions (X,Y,Z) of (3), namely

$$(1,2,5) \qquad (1,3,3) \qquad (2,2,2)$$

For k=2 we can accept half-integer solutions of (3), and so on. In effect this is a way of "stratifying" the rational solutions of an equation. It could probably be applied to other equations as well, i.e., given an equation with finitely many positive integer solutions, count the numbers of rational solutions whose common denominators divide k=1,2,3,...

Still another approach is to lower our ambitions and determine the number of 2-part unit fraction partitions of $1/k$. Let $U2(N)$

Unit Fraction Partitions

denote the number of distinct partitions of $1/N$ into exactly two unit fractions (not counting order). The first several values of $U2(N)$ for $N = 1, 2, 3,...$ are

1 2 2 3 2 5 2 4 3 5 2 8 2 5 5 5 2 8 ...

Surrisingly, this sequence doesn't appear in Sloane's Encyclopedia, so maybe it's worth covering this admittedly simple case. Clearly $U2(N)$ depends only on the exponents in the prime factorization of N.

In general, if N has the factorization

$$(p_1^{a_1})(p_2^{a_2})...(p_t^{a_t})$$

then

$$U2(N) = \frac{(2a_1 + 1)(2a_2 + 1) ... (2a_t + 1) + 1}{2}$$

Incidentally, if N is square-free the above formula reduces to $(3^t + 1)/2$. This sequence of numbers (1,2,5,14,41,122,365,...) DOES appear in Sloane's Encyclopedia, although I don't know if the reference relates this sequence to the numbers of partitions of $1/N$ into two unit fractions for square-free integers N with t prime divisors.

The function $U2(N)$ is formally similar to the number-of-divisors function $tau(N)$, although $U2$ is not multiplicative. We could generalize this to give the family of functions

$$F\_j(N) = \frac{(j a_1 + 1)(j a_2 + 1)...(j a_t + 1) + (j-1)}{j}$$

We know that $F\_1(N)$ gives the number of divisors of N, and $F\_2(N)$ gives the number of 2-part unit fraction partitions of N. I'm not sure what, if anything, is counted by $F\_3(N)$.

36

Reflective and Cyclic Sets of Primes

If the binary representations of the first several primes are reversed, the resulting number is also irreducible (i.e., either a prime or 1).

This was pointed out by John Rickert, who cited the following reflective pairs of primes

prime			reverse prime (or unit)		
2	10	01	1	(or 010 to 010)
3	11	11	3	
5	101	101	5	
7	111	111	7	
11	1011	1101	13	
13	1101	1011	11	
17	10001	10001	17	

The first prime whose binary reversal has a proper factorization is 19 = 10011, which gives 11001=25. Still, of the primes less than 2048, more than half of them have prime reflective partners.

I suspect the proportion of such primes becomes arbitrarily small as we go to larger and larger numbers. To illustrate,

Reflective and Cyclic Sets of Primes

here's a little table showing the number of such primes less than x for various values of x:

x	total number of primes < x	number of primes < x that give primes under binary reversal	ratio
100	25	21	0.8400
1000	168	102	0.6071
10000	1229	509	0.4141
100000	9592	3054	0.3184
1000000	78498	20054	0.2555
10000000	664579	141772	0.2133

I would expect that the fraction of primes whose reflective pairs are also primes is roughly what would be expected based on a random selection of an odd number with the correct number of binary digits for each prime. Thus it becomes increasingly less likely as the density of primes (slowly) decreases. Also, there are presumably infinitely many such prime pairs, although I don't know how to prove it.

One interesting approach to this problem might be based on interpreting the binary representations as polynomials with reciprocal roots, as discussed in The Fundamental Theorem For Palindromic Polynomials.

By the way, here's some C code for counting reflective pairs:

```
#include <stdio.h>
#include <stdlib.h>
#include <math.h>
unsigned long ic,d,nmax, nn, n, i, imax, mm, kk, aa[30];
unsigned long ictot,ptot;
main ()
  nmax=1000;
  for(nn=3;nn<=nmax;nn=nn+2)
  ic=0;for(d=3;d<sqrt(nn+1);d=d+2)if(nn%d==0)ic=1;
  if(ic==0)
  for(i=0;i<31;i++)aa[i]=0;
    n=nn;i=0;
    while(n>0){aa[i]=n%2;n=(n-aa[i])/2;i++;}imax=i-1;
```

```
mm=0;for(i=0;i<=imax;i++)mm=2*mm+aa[i];
ic=0;for(d=3;d<sqrt(mm+1);d=d+2)if(mm%d==0)ic=1;
ictot=ictot+ic; ptot=ptot+1;
printf(" %lu %lu %lu %lu %lu \n",ptot,nn,mm,ic,ictot);
```

On a slightly related point, we can also consider "complete cyclic sets of primes", i.e., primes whose base-b expansions are primes under all the rotations of their digits. Obviously there are none in the base 2 other than repunits, i.e., primes of the form $2^p - 1$.

However, for any higher base there exist complete cyclical prime sets. Focusing on the base 10, some trivial examples of two-digit cyclic pairs of primes are

11	13	37	17	79
11	31	73	71	97

For slightly less trivial examples, we have the three-digit cyclical prime sets

113	197	199	337
311	719	919	733
131	971	991	373

Among four-digit decimal numbers the only cyclic sets of primes are

1193	3779
3119	9377
9311	7937
1931	7793

Likewise among the five-digit decimal numbers we have only the sets

11939	19937
91193	71993
39119	37199
93911	93719
19391	99371

Finally, among six-digit decimal numbers the only cyclic sets of primes are

193939	199933
919393	319993
391939	331999
939193	933199
393919	993319
939391	999331

There are no complete cyclic sets of decimal primes with 7 or 8 digits. The next known cycle set is the repunit with 19 digits, then the repunits with 23, 317, 1031 digits, and so on. If we exclude repunits, it is not known if there are any complete cyclic sets of primes in the base 10 with more than 6 digits.

Of course, we can also find cyclic sets of primes in other bases. For example, in the base 7 we have the set

11515
51151
15115
51511
15151

Also, whenever a repunit is a prime, it gives a degenerate cyclic set, such as the number 1111111 in the base 3 (which equals the prime 1093 in decimal).

37

Waring's Problem

Fermat stated, and Lagrange proved, that every number can be expressed as the sum of four squares, but what about about cubes, forth powers, etc.? This is called Waring's Problem, because in 1770 Waring stated (without proof) that every number is expressible as a sum of 4 squares, and as a sum of 9 cubes, and as a sum of 19 fourth powers, "and so on". In 1909 Hilbert gave the first proof that for each positive integer exponent n there is an integer g(n) such that every integer is a sum of at most g(n) non-negative nth powers. A special case is Lagrange's Four Square Theorem (1770), which asserts that g(2)=4.

Around 1912 Wieferich and Kempner proved that g(3)=9, but it wasn't until 1964 than Chen proved g(5)=37. In the mean time, several people developed the techniques leading to the following result for all exponents greater than or equal to 6. (Square brackets signify that we are to take just the integer part of the enclosed quantity.)

Let $3^n = q\, 2^n + r$ with $0 < r < 2^n$. (In other words, q is the quotient of $3^n / 2^n$, and r is the remainder.)

If $r+q <= 2^n$ then $g(n) = 2^n + [(3/2)^n] - 2$.

If $r+q > 2^n$ then

$$g(n) = \begin{cases} 2^n + [(3/2)^n] + [(4/3)^n] - 2 & \text{if } Q = 2^n \\ 2^n + [(3/2)^n] + [(4/3)^n] - 3 & \text{if } Q < 2^n \end{cases}$$

where $Q = q + (q+1)(4/3)^n$.

Actually, the complicated case of $r+q > 2^n$ is somewhat academic, because it's been verified that $r+q < 2^n$ for every $n < 200000$, and no value of n for which $r+q$ exceeds 2^n is known. However, the possibility has not been ruled out, so we have to carry along all that extra baggage. If you're only interested in exponents less than 200000 all you have to remember is $g(n) = 2^n + [(3/2)^n] - 2$.

By the way, you may have noticed that the above results don't cover $g(4)$. It wasn't until 1986 that Balasubramanian, Dress, and Deshouillers finally proved Waring's assertion $g(4)=19$ to be correct.

On a closely related topic, let $G(n)$ denote the smallest number of nth powers that can represent all but finitely many exceptions.

For example, even though $g(3)$ is 9, there are really only two numbers (23 and 239) that cannot be expressed as sums of 8 cubes.

It's also known that only finitely many numbers (of which the largest is probably 454) cannot be expressed as sums of 7 cubes, so $G(3)$ is certainly no greater than 7. It may be as low as 4, but no one knows for sure. On the other hand, it's known that $G(4)=16$.

38

Cyclic Divisibility

The integer 865281023607 happens to be a multiple of 111111, but in addition it can easily be verified that every cyclical rotation of these decimal digits is also a multiple of 111111, as shown below

```
865281023607 = 111111 * 7787537
786528102360 = 111111 * 7078760
078652810236 = 111111 *  707876
607865281023 = 111111 * 5470793
360786528102 = 111111 * 3247082
236078652810 = 111111 * 2124710
023607865281 = 111111 *  212471
102360786528 = 111111 *  921248
810236078652 = 111111 * 7292132
281023607865 = 111111 * 2529215
528102360786 = 111111 * 4752926
652810236078 = 111111 * 5875298
```

Furthermore, if we reverse the digits of these number, we find that each of them is again a multiple of 111111, as shown below

```
706320182568 = 111111 * 6356888
870632018256 = 111111 * 7835696
```

Cyclic Divisibility

$$
\begin{aligned}
687063201825 &= 111111 * 6183575 \\
568706320182 &= 111111 * 5118362 \\
256870632018 &= 111111 * 2311838 \\
825687063201 &= 111111 * 7431191 \\
182568706320 &= 111111 * 1643120 \\
018256870632 &= 111111 * 164312 \\
201825687063 &= 111111 * 1816433 \\
320182568706 &= 111111 * 2881646 \\
632018256870 &= 111111 * 5688170 \\
063201825687 &= 111111 * 568817
\end{aligned}
$$

Of course, since the number

$$10^{12} - 1 = 999999999999$$

is a multiple of 111111, it's clear that the 10's complement of each of the above numbers is also a multiple of 111111. In other words, the number 134718976392 AND its reversal AND all of their cyclic rotations are each multiples of 111111.

What's going on here? Well, if we rotate the number 134718976392 to the right by one digit, and write the digits as a sequence, we see that they represent a cycle of the Fibonacci recurrence

$s[n] = s[n-1] + s[n-2]$ reduced modulo 10:

$s[n]$: 2 1 3 4 7 11 18 29 47 76 123 199 322 521 843 1364 ...

2 1 3 4 7 1 8 9 7 6 3 9 2 1 3 4 ....

The period of this sequence is 12, so to express the repeating decimal expansion

0.213471897639 213471897639 213471897639 ...

as a ratio of integers we decompose it into a geometric series

$$213471897639 \left(\frac{1}{10^{12}} + \frac{1}{10^{24}} + \frac{1}{10^{36}} + \ldots \right)$$

$$= \frac{213471897639}{10^{12} - 1} = \frac{1921249 * 111111}{9000009 * 111111} = \frac{1921249}{9000009}$$

So, the fact that the sequence of digits gives a decimal number with a large common divisor with $10^{12} - 1$ results in a significant reduction in the fractional expression for the repeating decimal. We also note that this large common factor is $(10^6 - 1)/(3^2)$.

More generally, we know that the period of any recurrence of the Fibonacci sequence modulo 10 must be a divisor of the least common multiple of the fundamental periods modulo 2 and 5. The fundamental period modulo 2 is $2^2 - 1 = 3$, whereas the fundamental period modulo 5 is $5(5-1) = 20$ (because 5 divides the discriminant of the Fibonacci polynomial). Hence, the period of any Fibonacci sequence must be a divisor of 60.

Now, if j is a divisor of 60 there exists an integer G_j that divides $10^j - 1$ such that if $\{d0, d1, d2, ..., dj\}$ is any cycle of positive residues modulo 10 that satisfy the Fibonacci recurrence, then the integer

$$d0 + d1*10 + d2*10^2 + ... + dj*10^j$$

is divisible by G_j. This is why all the rotations of any such sequence automatically are divisible by the same G_j. Also, the reverse recurrence has the same properties, noting that we can write it as $s[n-2] = -s[n-1] + s[n]$ which has the reverse characteristic polynomial $x^2 + x - 1$, whose discriminant is the same as that of $x^2 - x - 1$. Of course, the divisibility of the 10's complement integers is obvious from the fact that G_k divides $10^k - 1$.

The above is really just a special case of a much more general phenomenon. If the discriminant of a 2nd-degree polynomial $f(x)$ is divisible by the odd prime p, then we can consider the sequences that satisfy the linear recurrence corresponding to that polynomial, and reduce the sequence modulo $m = p$ or $2p$. The period of any such reduced sequence must be a divisor of $p(p-1)$ or $3p(p-1)$ respectively. If j is any such integer period length, then there exists an integer G_j that divides $m^j - 1$ such that if $\{d0, d1, d2, ..., dj\}$ is any cycle of positive residues modulo 10 that satisfy the Fibonacci recurrence, then the integer

Cyclic Divisibility

$$d0 + d1^*m + d2^*m^{\wedge}2 + \ldots + dj^*m^{\wedge}j$$

is divisible by G_j.

For example, staying with the Fibonacci recurrence, suppose we evaluate sequences modulo 5, which means they must have periods dividing 20. Once such sequence is

1 2 3 0 3 3 1 4 0 4 4 3 2 0 2 2 4 1 0 1

Interpreting these as the digits of a number in the base 5, this equals the decimal integer

20520701511336 = $(2^{\wedge}3)(3^{\wedge}2)(13)(521)(42080281)$

We can easily verify that this number, and all its rotations, reflections, and 5s-complements, share a greatest common divisor of $(2^{\wedge}3)(3)(13)(521)$ with the number

$5^{\wedge}20 - 1$ = $(2^{\wedge}4)(3)(11)(13)(41)(71)(521)(9161)$

For another example, consider the polynomial $f(x) = x^{\wedge}2 - 4x - 9$, whose discriminant is $4^*13 = 52$. Every sequence of residues mod 13 that satisfies this recurrence must have a period dividing 156. One such sequence, with a period of 12, is

1 2 4 8 3 6 12 11 9 5 10 7

which gives an integer with the decimal equivalent

13984822237218 = (2)(3)(7)(41)(61)(157)(847997)

and we can verify that this numbers and all its rotations, reversals, and 13s-complements share a common divisor of (2)(3)(7)(61)(157) with $13^{\wedge}12 - 1$.

For just one more example, consider the same polynomial, but this time evaluate the recurrence modulo $2(13) = 26$. One solution sequence, having a period of 12, is

1 2 17 8 3 6 25 24 9 18 23 20

which gives an integer with the decimal equivalent of

$$76753544008735881 = (3\wedge 6)(7)(19)(31)(37)(149)(4632011)$$

This number, and all its rotations, reflections, and 13s-complements share a common factor of $(3\wedge 4)(7)(19)(31)(37)$ with $26\wedge 12 - 1$.

Incidentally, notice that $m\wedge 2k - 1$ splits as $(m\wedge k + 1)(m\wedge k - 1)$, and it appears that in most cases $G_{\{2k\}}$ is a divisor of $m\wedge k - 1$. However the exception is the case with the Fibonacci recurrence (mod 5), when $G_{\{20\}}$ contained factors from both $5\wedge 10 + 1$ and $5\wedge 10 - 1$. This is also the only one of our examples in which the period of the sequence we examined was equal to the full period.

It's also interesting to examine the numbers produced by taking every jth element of one of these sequences. If j divides the period of the sequence then obviously the new sequence has a period reduced by a factor of j. Such a sequence still shares a large common factor with $m\wedge k - 1$. If j is coprime to the period of the sequence, then the new sequence must be either the same as the original, or else the reversal of the original. For example, if we take every 3rd element of the sequence with period 20 (mod 5) from above, we have

$$1\ 0\ 1\ 4\ 2\ 2\ 0\ 2\ 3\ 4\ 4\ 0\ 4\ 1\ 3\ 3\ 0\ 3\ 2\ 1$$

which is simply the reversal of the original sequence.

39

Unit Fractions and Fibonacci

The total electrical conductance of two components conntected in parallel is simply the sum of the two individual conductances, so the total resistance R is the inverse of the sum of the inverses of the individual resistances r1 and r2. In other words, resistances in parallel combine according to the equation.

Suppose we wish to generate integer solutions of the harmonic equation

$$\frac{1}{x} + \frac{1}{y} = \frac{1}{z}$$

This can be regarded as a special case of a more general expansion related to the Fibonacci numbers. Let $s[j]$, $j=0,1,2,...$ be a sequence of integers that satisfy the recurrence $s[k] = s[k-1] + s[k-2]$ with arbitrary initial values $s[0]$ and $s[1]$. It can be shown that for any integers m,n with m>n we have

$$\frac{1}{s[n-1]\,s[n]} = \frac{1}{s[m]\,s[m+1]} + \sum_{j=n}^{m} \frac{1}{s[j-1]\,s^r j = 1]}$$

For example, setting $s[0]=s[1]=1$ and n=5, m=10 gives

$$\frac{1}{40} = \frac{1}{65} + \frac{1}{168} + \frac{1}{442} + \frac{1}{1155} + \frac{1}{3026} + \frac{1}{7920} + \frac{1}{12816}$$

In general, to expand $1/D$ into a sum of unit fractions, the method is to split D into two factors, $D = pq$. Then we can set $s[0]=p$ and $s[1]=q$ and generate the s sequences as follows

k	s[k]	s[k] s[k-1]	s[k] s[k-2]
0	p		
1	q	pq	
2	p+q	q(p+q)	p(p+q)
3	p+2q	(p+q)(p+2q)	q(p+2q)
4	2p+3q	(p+2q)(2p+3q)	(p+q)(2p+3q)
5	3p+5q	(2p+3q)(3p+5q)	(p+2q)(3p+5q)
6	5p+8q	(3p+5q)(5p+8q)	(2p+3q)(5p+8q)
7	8p+13q	(5p+8q)(8p+13q)	(3p+5q)(8p+13q)
	etc	etc	etc

We can now express $1/pq$ as the sum of the inverses of the numbers in the third column down to the mth row, plus the inverse of the mth number in the second column. Thus we have

$$\begin{aligned}1/pq &= 1/p(p+q) + 1/q(p+q) \\ &= 1/p(p+q) + 1/q(p+2q) + 1/(p+q)(p+2q) \\ &= 1/p(p+q) + 1/q(p+2q) + 1/(p+q)(2p+3q) \\ &\quad + 1/(p+2q)(2p+3q) \\ &\quad \text{etc.}\end{aligned}$$

Of course, we can let m in equation (1) go to infinity, giving the infinite unit fraction expansion

$$\frac{1}{s[n-1]\,s[n]} = \sum_{j=n}^{\inf} \frac{1}{s[j-1]\,s[j+1]} \quad (2)$$

Unit Fractions and Fibonacci

This can also be generalized to higher order recurrences. For example, if we define the sequence s[j] to satisfy the 3rd order recurrence s[k] = s[k-2] + s[k-3] with the initial values a,b,c, then we can generate the following sequences

k	s[k]	s[k]s[k-1]s[k-2]	s[k]s[k-1]s[k-3]
0	a		
1	b		
2	c	abc	
3	a+b	bc(a+b)	ac(a+b)
4	b+c	c(a+b)(b+c)	b(a+b)(b+c)
5	a+b+c	(a+b)(b+c)(a+b+c)	c(b+c)(a+b+c)
6	a+2b+c	(b+c)(a+b+c)(a+2b+c)	(a+b)(a+b+c)(a+2b+c)
		etc	etc	etc

so we have

1/abc = 1/ac(a+b) + 1/bc(a+b)

= 1/ac(a+b) + 1/b(a+b)(b+c) + 1/c(a+b)(b+c)

= 1/ac(a+b) + 1/b(a+b)(b+c) + 1/c(b+c)(a+b+c) + 1/(a+b)(b+c)(a+b+c)

and so on. To illustrate, with a=3,b=7,c=11 this last formula gives

$$1/231 = 1/330 + 1/770$$

$$= 1/330 + 1/1260 + 1/1980$$

$$= 1/330 + 1/1260 + 1/3780 + 1/4158$$

and with a=23,b=c=1 it gives

$$1/23 = 1/24 + 1/552$$

$$= 1/48 + 1/50 + 1/552 + 1/1200$$

40

Solving Magic Squares

What's the most efficient way of generating all possible NxN magic squares? Obviously trying all $(N^2)!$ possible arrangements and checking for "magicness" would be prohibitive. Presumably this was not the method used by Bernard Frenicle de Bessy back in 1693 to determine that there are exactly 880 distinct 4x4 magic squares, not counting rotations and reflections. (Does anyone know what method he actually used?)

There are several more modern treatments of the subject, such as in the book "Winning Ways" by Conway, but they mainly seem to approach the question in terms of equivalence classes and transformations of various kinds. This approach is certainly illuminating, but it's also interesting to try just "solving" the problem algebraically.

In general, when dealing with NxN squares, where N is odd, it's helpful to subtract $(N^2 + 1)/2$ from each of the numbers 1 through N^2, since this makes all the common sums zero. For example, with a typical 3x3 square we subtract 5 from each number to give

```
a b c         1 -4  3
d e f   =     2  0 -2
g h i        -3  4 -1
```

which makes it immediately obvious that there is only one possible

Solving Magic Squares

magic square of order 3, up to rotations and reflections. This can also be seen by combining the algebraic conditions on the variables to eliminate all but two of the variables and arrive at the conic

$$2a^2 + 2ab + b^2 = 10 \qquad (1)$$

(See Note 1 for derivation.) This is an ellipse with the eight lattice points

$$(-3, 2) \quad (-3, 4) \quad (-1, 4) \quad (1, 2)$$
$$(3, -2) \quad (3, -4) \quad (1, -4) \quad (-1, -2)$$

each of which represents one of the eight rotations or reflections of a 3x3 magic square. Therefore, equation (1) and the eight lattice points can be regarded as the "solution" of 3x3 magic squares.

To explicitly solve for these lattice points, note that equation (1) was based on the linear conditions along with the sum of squares, and we can derive similar equations based on the sum of cubes, the sum of 4th powers, and so on. It turns out that the sum of cubes just gives equation (1) again, but the sum of 4th powers gives the quartic

$$6a^4 + 12a^3b + 10a^2b^2 + 4ab^3 + b^4$$
$$+ 96(2a^2 + 2ab + b^2) = 1078$$

Substituting for the quantity in parentheses from equation (1), this reduces to

$$6a^4 + 12a^3b + 10a^2b^2 + 4ab^3 + b^4 = 118 \qquad (2)$$

The loci of points (a,b) that satisfy equations 1 and 2 are shown in the figure below:

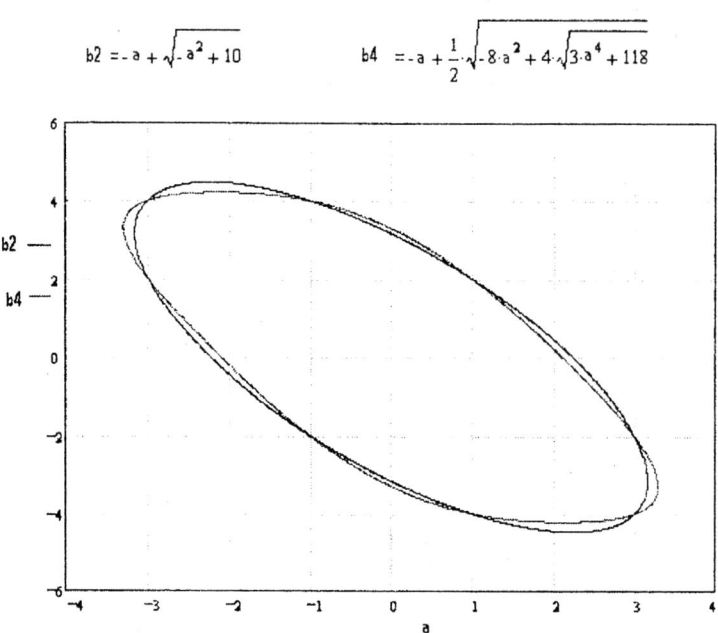

These two ovals intersect at the lattice points. The quartic (2) happens to have a nice solution in radicals, and the four roots are all similar to the one appearing in the figure above, with various signs. We can eliminate b from equations (1) and (2) by solving each of them for b and setting the results equal. Then with a little manipulation we arrive at the condition

$$a^4 - 10a^2 + 9 = (a^2 - 1)(a^2 - 9) = 0$$

which shows that the value of "a" (which appears in a corner of the 3x3 magic square) must be +-1 or +-3, and those are indeed the four corner values for the 3x3 magic square, and all the remaining entries are linearly related to these. Thus we have completely solved for the entries of a 3x3 magic square. Of course, there is essentially only one such square, so our solution is unique (up to rotations and reflections).

Solving Magic Squares

For a less trivial case, consider the general 4x4 magic square

$$\begin{array}{cccc} a & b & c & d \\ e & f & g & h \\ i & j & k & l \\ m & n & o & p \end{array}$$

For even-order squares the quantity $(N^2+1)/2$ is a half-integer, so in these cases we convert the original term x' in the range 1-16 to the corresponding centralized term x in the range $-(N^2-1)$ to $+(N^2-1)$ by applying the transformation $x = 2x'-(N^2+1)$. Thus, with 4x4 magic squares, instead of working with the values 1 through 16 we work with signed odd integers - 15,-13,..., +15.

Making use of the algebraic conditions on these 16 variables, we find that a square can often be determined by the values assigned to just 5 elements (but see below for exceptions). This is because, given the centralized values of g, h, k, l, and p, the values of f and o must satisfy the equation

$$(f+A)(Bo+C) = DE \qquad (2)$$

where

$A = 2p+k+l+h \qquad B = g+k \qquad D = A-g \qquad C = E+F$

$E = (l+k)(l+h) + p(g+h+l+k) \qquad F = (k+p)(k+g) + k(l+h)$

To illustrate, consider the square in Albrecht Durer's engraving "Melencolia", which has

$g'=11 \qquad h'=8 \qquad k'=7 \qquad l'=12 \qquad p'=1$

or, in centralized terms

$g=5 \qquad h=-1 \qquad k=-3 \qquad l=7 \qquad p=-15$

This gives A=-27, B=2, D=-32, E=-96, F=-54, and C=-150, so dividing the entire equation (2) by 2, the values of f and o must satisfy

$(27-f)(75-o) = 1536$

Translated back to the range 1-16, this condition is

$$(22-f')(46-o') = 384$$

which has just one solution in the range 1-16, namely [f'=10,o'=14], in agreement with Durer's engraving. From these it's a simple matter to fill in the remaining cells using relations such as a'+d'+p'+m'=34 and e+h = j+k, etc.

In a previous version of this note I asked if equation (2) necessarily leads to a unique magic square. Russell Blau has shown that it doesn't, by producing two distinct 4x4 magic squares with identical values of g, h, k, l, and p. In terms of centralized values these squares are

-5	-13	5	13		7	-13	-7	13
-3	9	-9	3		9	-3	-9	3
1	-11	11	-1		-11	1	11	-1
7	15	-7	-15		-5	15	5	-15

For both of these squares we have g=-9, h=3, k=11, l=-1, p=-15, and so A=-17, B=2, C=-26, D=-8, E=-40, F=14. Inserting these values into (2) and dividing by 2 gives the condition on f and o:

$$(f - 17)(o - 13) = 160$$

The solutions of this equation in odd integers in the range -15 to +15 are

$$(f,o) = (-3,5), (1,3), (7,-3), \text{ and } (9,-7)$$

but only (-3,5) and (9,-7) give valid magic squares. This naturally raises the question of how common this is, and whether there can be more than two solutions for a given set of parameters (g,h,k,l,p).

To answer these questions I checked all 524160 permutations of 16 numbers taken five at a time, and determined how many (if any) valid solutions they each give. It turns out that 4232 of those sets of five elements leads to one or more valid magic squares, as summarized in the table below:

Solving Magic Squares

k	Sets of Five Numbers Leading to k Valid Magic Squares	Number of Distinct Magic Squares
1	2176	2176
2	1656	3312
3	80	240
4	304	1216
5	0	0
6	16	96
	4232	7040

This accounts for all 7040 of the 4x4 magic squares. Of course, each of these can be oriented in four ways, and flipped over for four more ways, so there are really just 7040/8 = 880 distinct 4x4 magic squares up to rotations and reflections. Incidentally, here's an example of SIX distinct squares for a single set of parameters $\{g,h,k,l,p\}$:

```
 1  -9  -7  15       1  -7  -9  15      -3 -13   1  15
13   3 -11  -5      13   3 -11  -5       9   7 -11  -5
-13 -3  11   5     -13  -3  11   5      -9  -7  11   5
-1   9   7 -15      -1   7   9 -15       3  13  -1 -15

-3   1 -13  15      -9  -7   1  15      -9   1  -7  15
 9   7 -11  -5       3  13 -11  -5       3  13 -11  -5
-9  -7  11   5      -3 -13  11   5      -3 -13  11   5
 3  -1  13 -15       9   7  -1 -15       9  -1   7 -15
```

This is a degenerate case, in the sense that it gives A=-19, D=-8, and B=C=E=F=0, so the relation between f and o is identically satisfied, and therefore is of no use.

A couple of unanswered questions are: What's the simplest equation that relates these five parameters (analagous to (1) for 3x3 squares)? Does every lattice point of that surface correspond to a rotation or reflection of the square?

41

Concordant Forms

Recently the question was raised as to which primes are expressible in the form $(x^2 - 1)(y^2 - 1)$ for rational values of x and y. This is essentially equivalent to a question on "concordant forms", because if there are integers a,b,c,d such that

$$\left[\left\langle\frac{a}{b}\right\rangle^2 - 1\right]\left[\left\langle\frac{c}{d}\right\rangle^2 - 1\right] = p \qquad (1)$$

where p is a prime and the fractions are in lowest terms, then we have (a,b)=(c,d)=1 and without loss of generality

$$a^2 - b^2 = pd^2 \qquad c^2 - d^2 = b^2 \qquad (2)$$

Thus, the prime p is expressible in the form $(x^2 - 1)(y^2 - 1)$ if and only if the two equations

$$b^2 + d^2 = c^2 \qquad b^2 + pd^2 = a^2 \qquad (4)$$

are concordant, i.e., can be solved simultaneously in integers.

Andre Weil's "Number Theory, an Approach Through History" includes the following comments on concordant forms

"In 1777 and again in 1782 he [Euler] seeks criteria for...a 'double equation'

Concordant Forms

$$X^2 + Y^2 = Z^2 \qquad X^2 + NY^2 = T^2 \qquad (5)$$

to have infinitely many solutions. Not surprisingly, he fail (it is still an open problem)... His interest and our own turns to proofs of impossibility for the cases [N=3 or 4] and others equivalent to these two. One may well admire his beautiful technique (rather different from Fermat's) for applying the infinite descent to such problems."

It's interesting to see that (at least as of 1984) some aspects of this were still "open", although Weil doesn't specify which questions remain unanswered. Interestingly, Dickson's "History..." contains a list that supposedly contains all the integers N < 100 for which equations (5) are concordant, but the list excludes 47, 53, and 83, which are definitely concordant, as was shown by the solutions

$$(14663)^2 + (111384)^2 = (112345)^2$$
$$(14663)^2 + 47(111384)^2 = (763751)^2$$

$$(1141)^2 + (13260)^2 = (13309)^2$$
$$(1141)^2 + 53(13260)^2 = (96541)^2$$

$$(2873161)^2 + (2401080)^2 = (3744361)^2$$
$$(2873161)^2 + 83(2401080)^2 = (22062761)^2$$

Anyway, I've developed a proof of the impossibility of (5) for any prime p such that p is congruent to 3, 5, 9, 11, or 13 (mod 16) and every odd prime divisor of p-1 is congruent to 3 (mod 4). This is actually a fairly strong proposition; every prime less than 100 is either known to be concordant or is ruled out by this proposition.

Here's the proof. Beginning with equations (4)

$$b^2 + d^2 = c^2 \qquad b^2 + pd^2 = a^2 \qquad (4)$$

we can clearly assign any convenient signs to a,b,c,d because they only appear squared. Also, we know that d is even and therefore a,b,c are all odd, and we have (b,c)=(b,d)=(a,d)=1. Combining equations (4) gives

$$a^2 + (p-1)b^2 = pc^2 \qquad (6)$$

For some choice of signs for a,b,c there necessarily exists an integer k and positive mutually coprime integers u,v such that

$$\begin{aligned} ka &= u^2 - 2(p-1)uv - (p-1)v^2 \\ kb &= u^2 + 2uv - (p-1)v^2 \\ kc &= u^2 + (p-1)v^2 \end{aligned} \qquad (7)$$

This is just the parametric solution of (6), analagous to the well-known formulas for Pythagorean triples. To see that there must be a solution with u and v both positive, notice that k(b-a) = 2puv, so by giving k the same sign as (b-a) we can make 2puv positive. Also, notice that equation (6) modulo p is just (a+b)(a-b)=0, and since the signs of a,b are arbitrary, we can choose them such that (a-b) is divisible by p (and of course (a+b) is not).

It will be helpful later to have some bounds on the magnitude of k, so let's solve equations (7) for the three unknowns u^2, uv, v^2, to give

$$u^2 = k\frac{p(b+c) - (b-a)}{2p} \qquad uv = k\frac{b-a}{2p}$$

$$v^2 = k\frac{p(c-b) + (b-a)}{2p\,(p-1)}$$

Since a,b,c,p are all odd, and (b-a) is divisible by p, it follows that the denominator factors 2p are "absorbed" by the numerators.

The result is that k must divide each of the quantities

$$u^2 \qquad uv \qquad (p-1)v^2$$

Since u,v are coprime this implies that k must be a divisor of both u and (p-1). From this it also follows that (u+v) cannot be divisible by p, because then according to equations (7) both ka and kb would be divisible by p, and so would k(a+b), but since p doesn't divide a+b it would have to divide k, which is impossible because k divides p-1.

Concordant Forms

Now, combining expressions (7) for kb and kc, we find that the positive coprime integers u,v satisfy the equation

$$(k^2)(c^2 - b^2) = 4uv(u+v)(pv-(v+u)) = (kd)^2 \qquad (8)$$

It's clear that v is necessarily coprime to each of u, u+v, and (p-1)v-u, so v is always a square. Also, since (u+v) is prime to p, we know that (u+v) is co-prime to all the other factors, so it too must be a square. Thus we have coprime integers r,s such that $v=r^2$ and $(u+v)=s^2$. Substituting these values into (8) gives

$$uv(u+v)(pv-(u+v)) = (r^2)(s^2 - r^2)(s^2)(pr^2 - s^2)$$

so the condition reduces to

$$(s^2 - r^2)(pr^2 - s^2) = \text{square} \qquad (9)$$

Letting g denote the gcd of $(s^2 - r^2)$ and $(pr^2 - s^2)$. It follows that g is the greatest common divisor of u and (p-1). Remember we showed previously that k divides gcd(u,p-1), so k divides g. Anyway, we have coprime integers t,w such that

$$s^2 - r^2 = gt^2$$
$$pr^2 - s^2 = gw^2$$

Combining these two equations gives

$$(p-1)r^2 = g(t^2 + w^2)$$
$$(p-1)s^2 = g(pt^2 + w^2)$$

Since g divides (p-1) we have integers h,f such that $(p-1)/g = fh^2$ where f is squarefree, and so we arrive at

$$w^2 + t^2 = f(hr)^2 \qquad (10)$$
$$w^2 + pt^2 = f(hs)^2 \qquad (11)$$

Our objective is to show that if p = 3, 5, 9, 11, or 13 (mod 16) and all the odd prime divisors of (p-1) are congruent to 3 (mod 4), then these two equations are impossible. There are three cases to consider.

First, suppose f is divisible by one of the odd prime factors of (p-1). In this case equation (10) by itself is impossible by Fermat's theorem on sums of two squares, i.e., in the factorization of any sum of two squares, any prime congruent to 3 (mod 4) must occur with an even exponent. Thus, we can exclude any odd prime factors in f.

Next we consider the case f=2. Recalling that p is one of 3,5,9,11,13 (mod 16) and that the parameters t,w are coprime, it's easy to see that equations (10) and (11) are impossible (mod 16).

This leaves only the third possibility, namely, f=1. However, notice that if f=1 then equations (10) and (11) are identical to the original equations

$$b^2 + d^2 = c^2$$
$$b^2 + pd^2 = a^2 \qquad (4)$$

For any solution (a,b,c,d) of this pair of equations, define the "norm" of the solution as $(bd)^2$. In view of equation (8), the norm of our new solution (hs,w,hr,t) is given by

$$(wt)^2 = (s^2 - r^2)(pr^2 - s^2)/g^2$$
$$= (k^2 d^2) / (4 r^2 s^2 g^2)$$

But recall that k divides g, so putting g=km we have

$$(wt)^2 = (d^2)/(2rsm)^2$$

which shows that $(wt)^2$ is smaller than d^2, so it is obviously smaller than $(bd)^2$. Thus, the norm of our new solution is strictly smaller than the norm of the original solution. This implies an infinite sequence of integer norms with strictly decreasing magnitudes, which is impossible. This completes the proof.

I wrote a little program to search for such r,s for each prime p. The smallest examples for primes p < 100 are listed below.

Concordant Forms

p	r	s	g	v	u	v+u
7	1	2	3	1	3	4
11	1	3	2	1	8	9
17	1	3	8	1	8	9
23	5	6	11	25	11	36
31	1	2	15	1	3	4
41	1	3	8	1	8	9
47	13	36	23	169	1127	1296
53	5	13	4	25	144	169
59	1	3	2	1	8	9
61	1	7	12	1	48	49
71	1	6	35	1	35	36
79	1	2	3	1	3	4
83	17	33	2	289	800	1089
97	1	7	48	1	48	49

These results confirm that there are no solutions for primes p such that p=3,5,9,11,13 (mod 16) and all odd divisors of (p-1) are congruent to 3 (mod 4). (In fact, I've checked much further, but not found any counter-examples). For most other primes it's not hard to find a solution.

David Einstein and Allan MacLeod checked the primes less than 1000 and found the following results

104 Concordant primes (solutions found)
48 Discordant (no solutions; also excluded by theorem)
16 Unknown (not excluded by theorem, but no solution found)

As both David and Allan have pointed out, all 16 of these can be ruled out on the basis of the Birch-Swinnerton-Dyer conjecture, so there's probably not much point in searching for solutions for any of these. However, I'm still interested in finding an elementary proof that these are discordant. If anyone can find a solution for one of these, or explain why a solution is impossible, I'd be interested to hear about it.

42

Numbers Expressible As $(a^2 - 1)(b^2 - 1)$

Some time ago I mentioned that the number 588107520 is expressible in the form $(X^2 - 1)(Y^2 - 1)$ (where X,Y are integers) in five distinct ways, and asked if anyone knew a 6-way expressible number.

So far, no 6-way expressible number has been found, although such a number has not been proved impossible.

Regarding 5-way numbers, Dean Hickerson and Fred Helenius both independently found five more, so as of now the complete list of 5-way expressible numbers is

> 588107520
> 67270694400
> 546939993600
> 2128050512640
> 37400697734400
> 556606791861200

No one seems to know if there are infinitely many such numbers, or even if there are any more beyond this list.

There are several possible approaches to constructing numbers of this kind. One way is to notice that if a,b,c,d are four integers such that the product of any two of them is a "shy square", i.e.,

Numbers Expressible As $(a^2 - 1)(b^2 - 1)$

$$ab = x^2 - 1 \quad ac = y^2 - 1 \quad ad = z^2 - 1$$
$$bc = u^2 - 1 \quad bd = v^2 - 1 \quad cd = w^2 - 1 \qquad (1)$$

then clearly (abcd) can be expressed in at least three ways, namely

$$abcd = (ab)(cd) = (ac)(bd) = (ad)(bc) \qquad (2)$$

There are some nice parametric formulas for sets of four numbers with property (1). For example, take

$$a = n$$
$$b = q(qn+2)$$
$$c = (q+1)((q+1)n+2)$$
$$d = 4abc + 2(a+b+c) \qquad (3)$$

where n is an integer and q is any rational number such that b and c are integers. The product abcd is

$$F(q,n) = 4qn\,(qn+1)\,(qn+2)\,(q+1)\,(qn+n+1)\,(qn+n+2)\,(qn\,(q+1) + 2q+1)$$

The values of $F(q,n)$ are sure to have at least three representations, but they may have more, and obviously any number that has more than three must have three, so these are good numbers to check. In fact we find that each of the numbers

F(3,4)	=	588107520
F(5,5)	=	67270694400
F(10,3)	=	546939993600
F(12,2)	=	2128050512640
F(4,-20)	=	37400697734400
F(4/3,156)	=	5566067918611200

has five distinct representations, as mentioned previously. Since each of these numbers splits into two shy squares in five ways, each of them is 3-way expressible in 10 different ways (i.e., there are 10 ways of choosing 3 out of 5 factorizations). The fact that each of these numbers is of the form $F(q,n)$ implies that at least one of the 10 3-way sets must be of the form (2).

Not all 3-way solutions are of the form (2). The most general 3-way set would consist of eight components a,b,c,d,e,f,g,h, with the three factorizations

$$abcdefgh = (abcd)(efgh) = (abef)(cdgh) = (aceg)(bdfh) \quad (4)$$

Similarly, the most general 4-way set would consist of 16 components, and the most general 5-way set would consist of 32 components. To see why an N-way set may require 2^N components, notice that the components are bisected N times, and in each bisection a component either is or isn't in the same segment as, say, component "a".

Thus we can encode each component's behavior in an N-bit binarynumber, where of course a={111...1}. Every other N-bit number can occur, so there may be 2^N different types of components.

Anyway, another approach to constructing multi-expressible numbers is to look at integers of the form c^2-1 that can be expressed as a product of two other integers of the same form. Define

$$f(a,b,...,z) = (a^2 - 1)(b^2 - 1)...(z^2 - 1)$$

and then find three integers a,b,c such that f(a,b), f(a,c) and f(b,c) are all shy squares. Then the integer

$$N = f(a,b,c)$$

has the three representations

$$N = f(a,b)f(c) = f(a,c)f(b) = f(b,c)f(a)$$

Noticed that for any integer u the integers v such that f(u,v) is a shy square occur in a second-order linear recurring sequence.

For example, the integers v such that f(5,v) is a shy square are

..., 3821, 386, 39, 4, 1, 6, 59, 584, 5781, ...

where the terms satisfy the recurrence s[n] = 10*s[n-1] - s[n-2].

Numbers Expressible As $(a^2 - 1)(b^2 - 1)$

In general, for any integer u, the sequence has the central values

$$\ldots, u-1, 1, u+1, \ldots$$

and it satisfies the recurrence $s[n] = 2u*s[n-1] - s[n-2]$. Also, notice that for any integers u and j the integer $f(s[u;j], s[u;j+1])$ is a shy square. Therefore, the three numbers u, $s[u;j]$, and $s[u;j+1]$ satisfy the stated condition, so the number

$$N = h(u, s[u;j], s[u;j+1])$$

has three distinct representations as a product of two shy squares for any choice of the integers u and j. To illustrate, with u=5 and j=2 we have

$$s[5;2]=59 \qquad s[5;3]=584$$

so the number f(5,59,584) is expressible as a product of two shy squares in three distinct ways, as follows

$$f(5,59,584) = f(5,34451) = f(59,2861) = f(584,289)$$

Thus, the s-sequences together constitute a 2-parameter family of 3-way expressible numbers.

Furthermore, for any given integer u you can generate a sequence of this type from any integer 'a' such that

$$(u^2-1)(a^2-1) = (x^2-1) \qquad (5)$$

because if we define $b = x + au$ we have

$$(u^2-1)(b^2-1) = (y^2-1)$$

where $y = bu - a$. Repeating this process we can define $c = y + bu$ and then we have

$$(u^2-1)(c^2-1) = (z^2-1)$$

where $z = cu - b$, and so on. Substituting $y = bu - a$ into the equation $c = y + bu$ gives the recurrence

$$c = 2ub - a$$

so the numbers a,b,c,... constitute a sequence of solutions. To prove that $f(s[j],s[j+1])$ is itself a shy square for any of these s-sequences, notice that putting $x = b - au$ in (5) and expanding the terms gives

$$(ua)^2 - u^2 - a^2 + 1 = b^2 - 2abu + (au)^2 - 1$$

Adding $(b^2-u^2)(a^2-1)$ to both sides gives

$$(ab)^2 - a^2 - b^2 + 1 = (ab)^2 - 2abu + u^2 - 1$$

which factors as

$$(a^2-1)(b^2-1) = (ab-u)^2 - 1$$

showing that any two consecutive elements of the sequence a,b,c,... satisfy a relation of this form.

Yet another approach was taken by Dean Hickerson and (independently) Fred Helenius:

Define $\quad f(a,b) = (a^2 - 1)(b^2 - 1)$

and

$$g(m,n) = \frac{4mn(m^2 - 1)(n^2 - 1)(m n - 1)}{(m-n)^2}$$

If m and n are integers such that n-m divides m^2-1, then $g(m,n)$ is an integer and has three representations:

$$g(m,n) = f\left(m, \frac{2mn^2 - m - n}{n - m}\right) = f\left(n, \frac{2nm^2 - m - n}{n - m}\right)$$

$$= f\left(\frac{mn - 1}{n - m}, 2mn - 1\right)$$

Furthermore, for certain values of m and n there are additional representations. In particular, if we set $n = m+3$ where m is

Numbers Expressible As $(a^2 - 1)(b^2 - 1)$

not divisible by 3, we have a four-way expressible integer

$$g(m, m+3) = f\left(m, \frac{2m^3 + 12m^2 + 16m - 3}{3}\right)$$

$$= f\left(m+3, \frac{2m^3 + 6m^2 - 2m - 3}{3}\right)$$

$$= f\left(\frac{m^2 + 3m - 1}{3}, 2m^2 + 6m - 1\right)$$

$$= f\left(\frac{m^2 + 6m - 5}{3}, m^2 + 3m + 1\right)$$

This proves there are infinitely many four-way expressible numbers. By examining lots of these you sometimes find one that has a 5th representation. For example, with m=31 or 37 we have

g(31,34) = 546939993600 = f(31, 23869) = f(34, 21761)
                       = f(271, 2729)
                       = f(351, 2107) = f(701, 1055)

g(37,40) = 2128050512640 = f(9, 163097)
        = f(37, 39441)   = f(40, 36481)
        = f(493, 2959)   = f(985, 1481)

Another interesting case is to set $m = 2r^2 - r - 2$ and $n = 2r^2 - 1$, which gives the four-way expressible numbers

g(m,n) = f($2r^2 - r - 2$, $16r^5 - 24r^4 - 8r^3 + 16r^2 - 1$)

       = f($2r^2 - 1$, $16r^5 - 32r^4 - 4r^3 + 28r^2 - 2r - 5$)

       = f($(2r - 1)(2r^2 - 2r - 1)$, $(2r + 1)(4r^3 - 4r^2 - 4r - 3)$)

       = f($4r^3 - 2r^2 - 4r + 1$, $8r^4 - 12r^3 - 4r^2 + 6r + 1$)

Examining some of these reveals that with r = 3 or 4 there's a 5'th representation as well:

g(13,17) = 588107520 = f(13, 1871) = f(17, 1429) = f(55, 441)
         = f(79, 307) = f(129, 188)

g(26,31) = 67270694400 = f(26, 9983) = f(31, 8371)
         = f(161, 1611)
         = f(209, 1241) = f(433, 599)

Another family of 4-way expressible numbers is given by setting

$$m = r(r^2 - 3)/2 \qquad n = (r^3 + r^2 - 4r - 2)/2$$

We also have the miscellaneous result

g(209,365) =   F(4/3,156)   =   5566067918611200

which doesn't seem to be part of an algebraic family of 4-way numbers.

In summary, the basic building block of multi-representable numbers is the 2-parameter formula for three-way expressible numbers, which can be algebraically specialized to 1-parameter families of four-way expressible numbers. Some of these four-way expressibles also have a fifth representation, but it isn't clear whether there is an algebraic specialization to give these, or they are just numerical accidents.

There also appear to be some 5-way numbers that are not members of an algebraic 4-way family. It's unknown if there exist any 6-way expressible numbers.

43

Euclidean Algorithm

Suppose we wish to determine integers x,y such that

$$37x + 20y = 1209$$

Equations of this type are usually solved using what's called the Euclidean Algorithm, but a bare description of this algorithm in terms of modulo arithmetic is hard to follow for some students, mainly (I think) because they don't see the motivation behind each of the steps, so it just seems like a big recipe that happens to give the right answer. Let me see if I can describe the solution procedure in the most primitive terms, probably similar to the way in which Euclid's Algorithm was devised in the first place.

Given the equation $37x + 20y = 1209$, collect as many multiples of 20 as you can from each of the terms. This gives

$$20(x+y-60) + 17x = 9$$

Let's call the number in parentheses a, so we have a = x+y-60 and we can write the equation

$$20a + 17x = 9$$

Notice that this is an equation of the same form as the original, but with smaller coefficients. Let's repeat the procedure, this time collecting as many multiples of 17 as we can from each of the terms. The result is

$$17(a+x) + 3a = 9$$

Now let's call the number in parentheses $b = a+x$, so we have the equation
$$17b + 3a = 9$$

Repeating the procedure, this time we collect as many multiples of 3 as we can, giving

$$3(5b+a-3) + 2b = 0$$

Letting c denote the number in parentheses $c = 5b+a-3$ we have
$$3c + 2b = 0$$

Now repeat the procedure one more time, collecting multples of 2, to give
$$2(c+b) + c = 0$$

so if we set $d = c+b$ we have the equation $2d + c = 0$. At last we have an equation where one of the coefficients is 1, so we can say $c = -2d$. Now remember how we defined all of our parameters a,b,c,d

$$\begin{aligned} d &= c+b \\ c &= 5b+a-3 \\ b &= a+x \\ a &= x+y-60 \end{aligned}$$

We can substitute $c=-2d$ into the first of these equations to give $b = d - (-2d) = 3d$. Then we can substitute $c=-2d$ and $b=3d$ into the second equation to give
$$a = (-2d) - 5(3d) + 3 = -17d + 3$$

then we can substitute $b=3d$ and $a=-17d+3$ into the 3rd equation to give
$$x = (3d) - (-17d + 3) = 20d - 3$$

Euclidean Algorithm

Now if we substitute x=20d-3 and a=-17d+3 into the 4th equation we find

$$y = (-17+3) - (20d-3) + 60 = -37d + 66$$

Therefore, we have shown that the original equation is satisfied for integers x,y if and only if

$$x = 20d - 3$$
$$y = -37d + 66$$

where d is any integer. We could tidy this up by setting k = d-1 to give the reduced form

$$x = 20(k+1) - 3 = 20k + 17$$
$$y = -37(k+1) + 66 = -37k + 29$$

where k is any integer. Thus there are infinitely many solutions, the smallest of which is x=17, y=29. In general, if A,B,C have no common factors, the solution of Ax+By=C is of the form

$$x = Bk + x0$$
$$y = -Ak + y0$$

where x0,y0 is a particular solution, whose values can be derived by the Euclidean Algorithm as illustrated above.

Incidentally, if you're comfortable with modulo arithmetic, notice that the equation 37x + 20y = 1209 immediately implies

$$x = 9/17 \pmod{20} \qquad y = 5/4 \pmod{37}$$

so the solution is essentially equivalent to modulo division.

44

On the Density of Some Exceptional Primes

As discussed in the note On Case 1 of Fermat's Last Theorem, the congruence

$$(x+1)^p - (x)^p - 1 = 0 \pmod{p^2} \qquad (1)$$

must have a non-trivial solution x coprime to p in order for there to exist an integer solution of Fermat's equation $x^p + y^p + z^p = 0$ with xyz coprime to p. (Incidentally, the numbers 1093 and 3511 are the only known primes such that x=1 is a root of congruence (1)).

However, this congruence has at least one non-trivial root for every prime p of the form 3k+1. (Here non-trivial means other than x = 0 or -1 modulo p). On the other hand, there are no non-trivial solutions at all for most primes of the form 3k-1. The exceptions less than 7000 are listed below

```
  59   83  179  227  419  443  701  857  887  911  929  971  977
1091 1109 1193 1217 1223 1259 1283 1289 1439 1487 1493 1613 1637
1811 1847 1901 1997 2003 2087 2243 2423 2477 2579 2591 2729 2777
2969 3089 3137 3191 3203 3251 3467 3527 3533 3881 3917 3929 4001
4079 4091 4643 4649 4691 4889 4943 5189 5303 5351 5711 5843 5849
5903 6257 6359 6389 6551 6569 6581 6947 6959
```

On the Density of Some Exceptional Primes

The total numbers of primes of the form 3k-1, and the numbers of these "exceptional" primes, up to various levels are listed below

	Tot	Excpt	dens
1000	86	13	0.151
2000	153	30	0.196
3000	221	40	0.181
4000	277	51	0.184
5000	337	59	0.175
6000	397	66	0.166
7000	454	74	0.162

Clearly if x is a root of (1) for a given prime p, then so is x+kp for every k, so we can represent all the non-trivial roots by just those in the range 1 to p-2. Furthermore, if x is a root, then -(x+1) is also a root, in view of

$$(-(x+1)+1)^p - (-(x+1))^p = (-x)^p - (-(x+1))^p \pmod{p^2}$$

Therefore, we can represent all the non-trivial roots by the set of roots in the range from 1 to (p-1)/2.

The table below shows the number of primes of the form 3k-1 less than 160000, along with the number exceptional primes in this range.

range	# of primes the form 3k-1	number of exceptional primes	ratio
0 - 10000	616	96	0.15584
10000 - 20000	520	83	0.15961
20000 - 30000	497	83	0.16700
30000 - 40000	479	72	0.15031
40000 - 50000	463	66	0.14254
50000 - 60000	466	77	0.16523
60000 - 70000	449	74	0.16481
70000 - 80000	448	65	0.14508

80000 - 90000	436	71	0.16284
90000 - 100000	432	78	0.18055
100000 - 110000	427	66	0.15456
110000 - 120000	432	67	0.15509
120000 - 130000	440	73	0.16590
130000 - 140000	427	63	0.14754
140000 - 150000	408	64	0.15686
150000 - 160000	421	59	0.14014
0 - 160000	7361	1157	0.15717

It certainly appears that the density of exceptional primes among all primes of the form 3k-1 is fairly constant, independent of the size of the primes involved. It's also worth noting that whenever a prime of this form has (non-trivial) solutions, the number of solutions in the reduced range 1 to (p-1)/2 is a multiple of three (as explained below). The distribution of the number of solutions for the primes less than 160000 is shown below

# of primes p = 3k-1	# of solutions solutions
6204	0
1055	3
98	6
4	9

The density seems to drop roughly by factors of 1/6, 1/11, and 1/24 for each additional set of three roots, suggesting that primes with 12 roots would have a density of roughly 1/48 times the density of primes with 9 roots.

The reason that these roots occur in sets of 3 is easier to see if we consider the whole range of residues from 1 to p-2. Over this range the solutions of

$$(x+1)^p - x^p - 1 = 0 \pmod{p^2}$$

occur in sets of six related values, because if x is a solution, then another solution is given by replacing x with either -(1+x) as we saw previously, or with -(1 + 1/x), where the inverse is taken modulo p. The former has a period of 2, whereas the latter has a period of 3, and together these substitutions generate a group of order 6 with the members

$$a(x) = x \qquad b(x) = 1/x \qquad c(x) = -(1+x)$$

$$d(x) = -1/(1+x) \qquad e(x) = -x/(1+x) \qquad f(x) = -(1+x)/x$$

This is isomorphic to the group of symmetries of an equilateral triangle. The operation with period 2 is like flipping the triangle over in the plane, and the operation with period 3 is like rotating the triangle through 120 degrees. In general, these operations canbe used to partition the residues modulo p into equivalence classes of 6 elements that are related according to these operations. For example, with p=59 we get the following sets

a	b	c	d	e	f	
1	1	57	29	29	57	
2	30	56	39	19	28	
3	20	55	44	14	38	*
4	15	54	47	11	43	*
5	12	53	49	9	46	
6	10	52	42	16	48	
7	17	51	22	36	41	
8	37	50	13	45	21	
18	23	40	31	27	35	
24	32	34	33	25	26	

The two sets marked with asterisks are the values that satisfy (1) in the range from 1 to p-2. Obviously for each set we have the relations

$$a + c = d + e = b + f = -1$$
$$ab = cd = ef = 1 \qquad \pmod{p}$$

Notice that for primes congruent to +1 (mod 3) there is alway~ a degenerate set corresponding t~ the algebraic factor

$(1 + x + x^2)^2$ of equation (1) when p is of this form. (For more on this, see the note Sums of Powers in Terms of Symmetric Functions. For example, with p=31 we have the sets

a	b	c	d	e	f
1	1	29	15	15	29
2	16	28	10	20	14
3	21	27	23	7	9
4	8	26	6	24	22
5	25	25	5	25	5
11	17	19	18	12	13

Here we see that 5 and 25 are roots of $(1+x+x^2)^2 = 0$ (mod $31)^2$. Also, the set containing 1 always consists of just three distinct elements, +1, -2, and (p-1)/2. So, if p is of the form 6k+1, the residues from 1 to p-2 are partitioned into one set with 3 members, one set with 2 members, and then the remaining

$$(p-2)-3-2 = 6k+1-2-3-2 = 6(k-1)$$

residues are partitioned into k-1 sets. On the other hand, if p is of the form 6k-1, the algebraic factor $(1+x+x^2)$ occurs only to the 1st power, so there is no degenerate set of 2 elements.

Thus the residues are patitioned into one set of 3 elements, and then the remaining

$$(p-2)-3 = 6k-1-2-3 = 6(k-1)$$

residues are partitioned into k-1 sets. We recall that only the primes 1093 and 3511 have the degenerate set containing 1 as solutions of (1).

Just to give one more example of an exceptional prime, with p=83 we have the partition

a	b	c	d	e	f
1	1	81	41	41	81
2	42	80	55	27	40
3	28	79	62	20	54

4	21	78	33	49	61	
5	50	77	69	13	32	
6	14	76	71	11	68	
7	12	75	31	51	70	
8	52	74	46	36	30	*
9	37	73	58	24	45	
10	25	72	15	67	57	
16	26	66	39	43	56	
17	44	65	23	59	38	
18	60	64	48	34	22	
19	35	63	29	53	47	

Again the set marked with an asterisk represents the solutions of congruence (1) in the range 1 to p-2.

Regarding the density of these primes, notice that the expression $(x+1)^p - x^p - 1$ is always a multiple of p, so we can divide that out, and we are left with a polynomial that wish to set equal to zero modulo p (so that the full expression vanishes modulo p^2).

For any given value of x, if we assume the value of this polynomial is randomly distributed over the residues mod p, we can say that the chances are about 1/p that it will vanish.

Now, the number of independent x values that we can try is really only (p-5)/6 (which is k-1 from above), because the solutions cover the equivalence classes described above. Thus, for any given prime p we have (p-5)/6 attempts, each of which has a 1/p probability of success, so overall the probability of at least one solution (up to equivalence class) should be about

$$[(p-5)/6](1/p) = (1/6)(1 - 5/p)$$

Thus it isn't too surprising that the density is roughly 1/6 = 0.167.

It would be interesting to know if there is any other way of characterizing these exceptional primes, besides evaluating (1).

Double Equations from Triangles in Squares Given the square [1234] in the figure below, is it possible for both of the inscribed right triangles [125] and [345] to have integer (or rational) edge lengths?

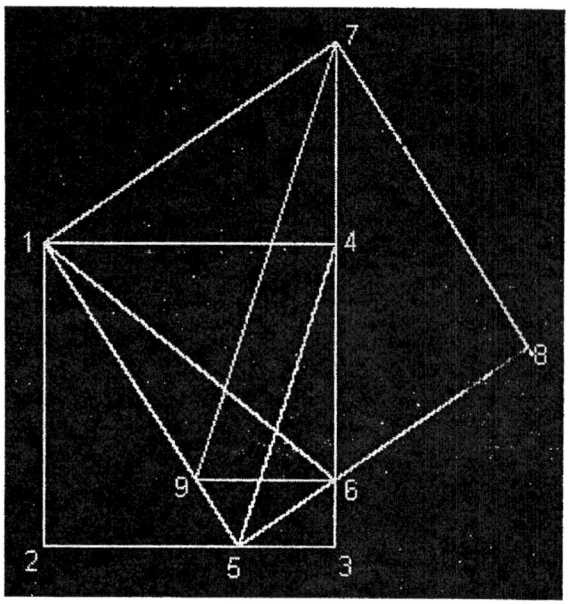

If we let "a" denote the length of the edge [12] and "b" denote the intermediate length [25], then obviously the condition for both triangles to be Pythagorean is that both of the equations

$$a^2 + b^2 = f^2 \qquad a^2 + (a-b)^2 = h^2 \qquad (*)$$

are satisfied, where f and h are integers representing the lengths of the "hypotenuses" [15] and [45] respectively.

Now suppose that instead of drawing the line from [5] to [4], we draw a line from [5] to [6] so that the angle (156) is a right angle, and then we draw the line [6,1]. Thus the original square [1234] is partitioned into four right triangles. Is it possible for all four of these to be Pythagorean triangles? Interestingly, the necessary and sufficient condition for this is the same as for the previous case, i.e., a (non-trivial) integer solution to the double equation (*).

To show this, let's assign letters to the segment lengths as follows:

$$[12] = a \quad [25] = b \quad [53] = c$$
$$[36] = d \quad [46] = e \quad [15] = f$$
$$[16] = g \quad [45] = h \quad [56] = j$$

By similar triangles we have $a/c = b/d = f/j$. Also, we can immediately express c,d,e in terms of a,b as follows

$$c = a-b$$
$$d = bc/a = b(a-b)/a$$
$$e = a-d = a - b(a-b)/a$$

Now, in order for all four of the right triangles partitioning the square to be rational (which can easily be converted to integers), we must have in addition to $a^2 + b^2 = f^2$ the equations

$$j^2 = c^2 + d^2 \qquad\qquad g^2 = a^2 + e^2$$

However, the triangle with edge lengths "cdj" is similar to "abf", so if the "abf" triangle is rational, then it follows that the "cdj" triangle is also rational. Specifically we have $j^2 = (cf/a)^2$.

Hence the only real requirement beyond the "abf" triangle being rational is that the "aeg" triangle be rational, which is true if and only if

$$a^4 + e^\wedge = + \left[a - \frac{b(a-b)}{a} \right]^2 = \frac{a^4 + (a^2 - ab + b^2)^2}{a^2}$$

is a rational square. Obviously the denominator is a square, so we need only consider the numerator, which factors as

$$a^4 + (a^2 - ab + b^2)^2 = [a^2 + b^2][a^2 + (a-b)^2]$$

The first factor on the right side is already known to be a rational square, since we have required that "abf" is a rational triangle.

Therefore, the other factor must also be a rational square, and so we arrive at the same double equation as (*) above.

Of course, the equivalence of the rationality conditions for triangles [345] and [156] was to be expected, because these two triangles are obviously similar (noting that f/j=a/c) and have at least one rational edge assuming that [125] is rational. As a result, we can construct a new square [1587] and we find that point 7 lies along the line {364}. Needless to say we have [678] similar to [512], and [679] is similar to [345] and [516].

But none of this answers the original question, which is whether such constructions are actually possible, i.e., whether there is an integer solution of the double equation

$$x^2 + y^2 = m^2 \qquad x^2 + (x-y)^2 = n^2$$

The question of whether two quadratic forms in two variables has solutions, and if so, whether it has infinitely many, has been studied for many years, going back to Diophantus, Bachet, Fermat, Euler, and so on. Notice that the right-hand equation can also be written in the form $2x^2 - 2xy + y^2 = n^2$. Many different techniques have been developed to tackle this kind of problem, but it still is not completely solved for arbitrary pairs of quadratic forms.

To tackle this particular pair of equations, we first note that any common factor in x and y can be divided out of both equations, so we can assume that both are primitive Pythagorean triples. From this it follows that x and y have opposite parity, as do x and x-y, which implies that x must be even and y must be odd. Consequently we have coprime integers A,B with opposite parity, and coprime integers C,D with opposite parity, such that

$$\begin{array}{ll} x = 2AB & x = 2CD \\ y = A^2 - B^2 & (x-y) = C^2 - D^2 \\ m = A^2 + B^2 & n = C^2 + D^2 \end{array}$$

This shows that AB = CD, so this product must have four pairwise coprime factors r,s,R,S (precisely one of which is even) such that

On the Density of Some Exceptional Primes

$$A = rs \quad B = RS \quad C = rR \quad D = sS$$

Adding the previous expressions for y and x-y gives

$$x = A^2 - B^2 + C^2 - D^2$$
$$= (rs)^2 - (RS)^2 + (rR)^2 - (sS)^2$$
$$= (r^2 - S^2)(s^2 + R^2)$$

Also, since $x = 2AB = 2CD = 2rsRS$, we have

$$(r^2 - S^2)(s^2 + R^2) = 2rsRS \qquad (1)$$

Since r,s,R,S are pairwise coprime, we know that both r and S are coprime to the first factor on the left, and both s and R are coprime to the second factor. Hence, depending on which of the two left hand factors is even (recalling that precisely one of r,s,R,S is even) we have one of two cases:

Case 1: Either r or S is even, and we have

$$r^2 - S^2 = Rs \qquad s^2 + R^2 = 2rS \qquad (2a,b)$$

In this case we can factor the left hand equation to give

$$(r+S)(r-S) = Rs$$

and since r,s,R,S are pairwise coprime we have pairwise coprime integers u,v,U,V such that

$$uv = r+S \qquad UV = r-S \qquad uV = R \qquad vU = s$$

Hence we have $r = (uv+UV)/2$ and $S = (uv-UV)/2$, and we can insert these expressions into (2b) to give

$$(vU)^2 + (uV)^2 = 2(uv+UV)(uv-UV)$$

Expanding the righthand product and re-arranging, we have

$$U^2 (v^2 + 2V^2) = u^2 (2v^2 - V^2) \qquad (3)$$

and re-arranging differently gives the alternate form

$$V^2 (u^2 + 2U^2) = v^2 (2u^2 - U^2) \qquad (4)$$

Since $\gcd(U,u)=1$ we know that u^2 divides $v^2 + 2V^2$, and so on.

Hence equation (3) can be written as

$$\frac{(v^2 + 2V^2)}{u^2} = \frac{(2v^2 - V^2)}{U^2} = M$$

for some positive integer M. Likewise equation (4) can be written in the form

$$\frac{(u^2 + 2U^2)}{v^2} = \frac{(2u^2 - U^2)}{V^2} = N$$

for some positive integer N. Consequently we have the equivalent pairs of equations

$$v^2 + 2V^2 = Mu^2 \qquad 2v^2 - V^2 = MU^2 \qquad (5a,b)$$

$$u^2 + 2U^2 = Nv^2 \qquad 2u^2 - U^2 = NV^2 \qquad (6a,b)$$

Multiplying (5a) by N and making the substitutions for Nv^2 and NV^2 from equations (6) gives

$$MNu^2 = Nv^2 + 2(NV^2) = (u^2 + 2U^2) + 2(2u^2 - U^2)$$
$$= 5u^2$$

This shows that $MN = 5$ for positive integers M,N, so either $M=1, N=5$ or else $M=5, N=1$. Both of these lead to the same reciprocal pair of quadratic forms (up to some permutation of the variables)

$$v^2 + 2V^2 = u^2 \qquad 2v^2 - V^2 = U^2$$

Case 2: Either R or s is even, and we have

$$r^2 - S^2 = 2Rs \qquad s^2 + R^2 = rS \qquad (7a,b)$$

In this case the left hand side is even, as is Rs, so we can write

$$\frac{r+S}{2} \cdot \frac{r-S}{2} = \frac{R}{2} \cdot s$$

assuming R is even. These factors are all coprime, so there are pairwise coprime integers u,v,U,V such that

$$uv = (r+S)/2 \qquad UV = (r-S)/2 \qquad uV = R/2 \qquad Uv = s$$

This implies $r = uv+UV$ and $S = uv-UV$. Inserting these into (7b) gives

$$(Uv)^2 + 4(uV)^2 = (uv+UV)(uv-UV)$$

Expanding and re-arranging gives

$$U^2(v^2 + V^2) = u^2(v^2 - 4V^2)$$

and

$$V^2(4u^2 + U^2) = v^2(u^2 - U^2)$$

Proceding in the same way as in Case 1, we find that this leads to a pair of equations of the form

$$v^2 + V^2 = u^2 \qquad v^2 - 4V^2 = U^2$$

In this case we see the question is equivalent to asking whether -4 is a "concordant number", defined as an integer N such that $x^2 + y^2$ and $x^2 + Ny^2$ can both be squares simultaneously. This is discussed at length in the note Concordant Forms, where a proof is given that a large class of positive prime values of N are not concordant.

A different approach is to cast the problem in the form of an elliptic curve. Returning to equation (1)

$$(r^2 - S^2)(s^2 + R^2) = 2rsRS$$

we see that the first factor on the left can be divided by rS and the second factor by Rs to give

$$[r/S - S/r][s/R + R/s] = 2$$

Thus we have rational numbers $X=r/S$ and $Y=R/s$ such that

$$\left\langle X - \frac{1}{X} \right\rangle \left\langle Y + \frac{1}{Y} \right\rangle = 2$$

Multiplying through by XY gives

$$(X^2 - 1)(Y^2 + 1) - 2XY = 0$$

If we define the new variable Z such that $X = (Z+Y)/(Z-Y)$, then Z is rational if X and Y are rational (and of course X is rational if Z and Y are rational). Notice that if $Y=0$ then $X=1$ for ANY value of Z, and this is a solution of the equation. Substituting for X gives

$$\frac{2Y(2ZY^2 + 2Z - Z^2 + Y^2)}{(Z-Y)^2} = 0$$

Setting aside the trivial solution at $Y=0$, and assuming Y is not equal to Z (which corresponds to infinite X) we are left with

$$2ZY^2 + 2Z - Z^2 + Y^2 = 0$$

which we can solve for Y^2 to give

$$Y^2 = \frac{Z(Z-2)}{(2Z+1)}$$

On the Density of Some Exceptional Primes

If we now define the variable W by the bi-rational form $Y = W/(2Z+1)$ and substitute for Y into the above expression, we get the elliptic curve

$$W^2 = Z(Z-2)(2Z+1)$$

This is nearly identical to the elliptic curve that arises when proving that there cannot exist four squares in arithmetic progression.

Only the sign in the second factor is different. As discussed in Weil's historical review of "Number Theory", essentially this same problem was treated by both Euler and Fermat.

A plot of the real part of the elliptic curve $y^2 = x(x-2)(2x+1)$ is shown below.

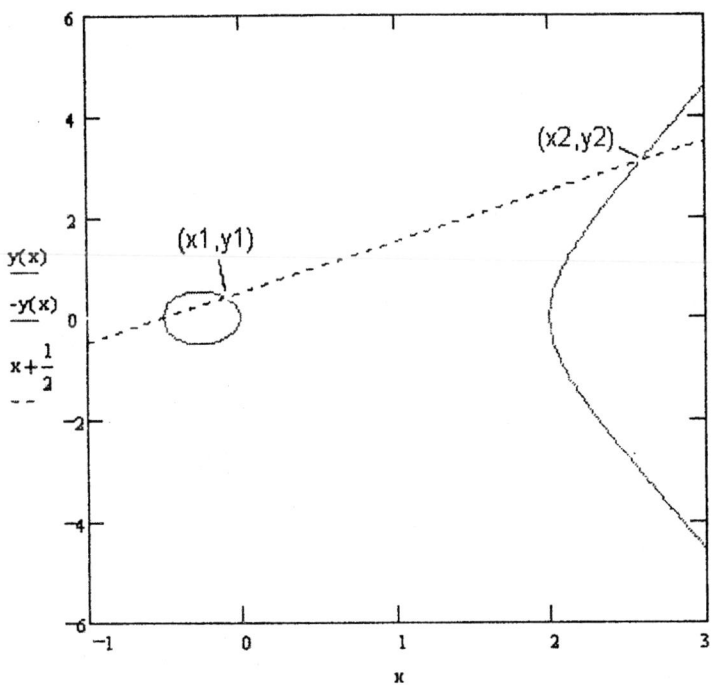

There are obviously at least three rational points on this curve, given by (x,y) = (0,0), (2,0), and (-1/2,0). Notice that these three rational points lie along the "horizontal" axis of symmetry of the curve. Any straight line passing through the closed loop on the left and striking the open branch of the curve on the right has three real points of intersection, and obviously if two of those points are rational then the third must be also. This shows how, if we are given any two rational points, we can generally construct a third, simply by drawing a line through the two given points and locating the remaining intersection point.

For example, suppose the point (x_1, y_1) on the curve above is a rational point. In that case we could draw the line through the two points $(-1/2, 0)$ and (x_1, y_1) as shown, and then we would have a new rational solution at the point (x_2, y_2). The equation of the line is

$$y = \frac{y_1}{x_1 + 1/2} x + \frac{y_1}{2x_1 + 1}$$

Squaring this and equating it with $y^2 = x(x-2)(2x+1)$ gives a cubic equation in x whose three roots are the x values of the three points of intersection between the line and the elliptic curve. We already know that two of these roots are $x = -1/2$ and $x = x_1$, so we can easily determine the third root and the corresponding value of y:

$$x_2 = -\frac{(y_1)^2}{x_1(2x_1 + 1)^2} \qquad y_2 = y_1 \frac{2x_2 + 1}{2x_1 + 1}$$

Hence if x_1 and y_2 are both rational, then so are x_2 and y_2. Once we have found this new rational point we can draw lines through it and any previously found rational points to generate still more rational solutions. Of course, in order to accomplish this we need first to have one rational point off the axis of symmetry.

45

Recurrences and Pell Equations

For any sequence s[0], s[1], s[2],... that satisfies the second-order linear recurrence

$$s[n] = A\, s[n-1] + B\, s[n-2]$$

with arbitrary constants A and B we have the algebraic identity

$$s[n]^2 - s[n-1]\, s[n+1] = [(\, s[1]^2 - s[0]\, s[2]\,)\,(-B)]^{n-1}$$

for all n. If B = 1, it follows that this quantity has constant magnitude with alternating sign. Denoting this quantity by D, we can express it's magnitude in terms of the initial values s[0] and s[1] as

$$|D| = |s[1]^2 - s[0]s[1] - s[0]^2|$$

For convenience, let n and m denote the initial values s[0] and s[1] respectively. Then we have a magnitude D corresponding to every pair of integers n,m such that $D = \pm(m^2 - mn - n^2)$. If we solve the quadratic equation

$$m^2 - nm - (n^2 \pm D) = 0$$

for m we have

$$m = \frac{n +- \sqrt{5n^2 + 4D}}{2}$$

(and the same for -D) which implies that the quantity in the square root must be a square, i.e., there must be an integer k such that $5n^2 + 4D = k^2$. Hence we seek solutions of the Pell equation

$$k^2 - 5n^2 = 4D$$

If D is a prime p, this equation signifies that $k^2 = 5n^2$ (mod p), which has solutions if and only if 5 is a square modulo p. Now, by quadratic reciprocity, this is the case if and only if p is a square modulo 5, which means that p is congruent to either +1 or -1 mod 5 (and therefore mod 10 as well). Thus all and only primes of the form 10j+1 and 10j-1 have solutions.

The first several solution pairs (n,m) for D = 1 are listed below. For each n there are two m values, given by the two roots of the quadratic.

```
s[0] =  n  =   1    3    8   21    55   144  377   987
s[1] = m1  =   2    5   13   34    89   233  610  1597
       m2  =  -1   -2   -5  -13   -34   -89 -233  -610
```

As we would expect, the soutions with D=-1 are essentially the same, but shifted by one index. Clearly these (n,m) solutions are all equivalent in the sense that they all lie within the same recurring sequence, i.e., the Fibonacci sequence. Therefore, we say there is essentially just one solution for D = +-1.

Similarly if D equals the "special" prime 5 (special because 5 is the discriminant of the characteristic polynomial of the Fibonacci recurrence, as shown by the 5 appearing in the Pell equation), there is essentially just one equivalence class of solutions, as shown below.

```
s[0] =  n  =   1    4   11    29    76   199   521
s[1] = m1  =   3    7   18    47   123   322   843
       m2  =  -2   -3   -7   -18   -47  -123  -322
```

However, if D is an odd prime congruent to +1 or -1 modulo 5, there are two equivalence classes of solutions, in the sense that the (n,m) solutions all lie within one of two distinct sequences. To illustrate, the first several (n,m) solutions for D=11 are listed below.

```
s[0] = n  =  1   2   5   7   14   19   37  50  97  131
s[1] =m1  =  4   5   9  12   23   31   60  81 157  212
       m2 = -3  -3  -4  -5   -9  -12  -23 -31 -60  -81
```

In this case the solution pairs (n,m) all lie in one of TWO distinct recurring sequences, namely

$$1 \quad 4 \quad 5 \quad 9 \quad 14 \quad 23 \quad 37 \quad 60 \quad 97 \quad 157 \ldots$$
or
$$2 \quad 5 \quad 7 \quad 12 \quad 19 \quad 31 \quad 50 \quad 81 \quad 131 \quad 212 \ldots$$

We could, however, regard this as just a single doubly-infinite sequence by proceding to negative indices (since we are assuming B=1). Tis leads to the single sequence

$$\ldots\ -19 \quad 12 \quad -7 \quad 5 \quad -2 \quad 3 \quad 1 \quad 4 \quad 5 \quad 9 \quad 14 \quad 23 \quad 37 \quad 60 \ldots$$

The sequences for the special values D=1 and D=5 are symmetrical about the origin, so they give only one distinct forward sequence, whereas for all other primes (congruent to +-1 mod 5) the sequences are not symmetrical, so they give two distinct forward sequences.

If we consider (square-free) composite values of D, it's clear that they must be products of the special numbers 1 and 5, and the primes congruent to +-1 modulo 5. These must include all the numbers given by $j^2 - j - 1$, meaning the sequence

-1 1 5 11 19 29 41 55 71 89 109 131 155 181 209...

The reason we know solutions exist for all these values is because if we solve the equation

$$j^2 - j - (1+D) = 0$$

for j we give the same Pell equation as before, with n=1. However, some of the D values with solutions (such as D=31) do not have a solution with n=1, so those are not given by this function. If a D value has a solution with n=2, it is in the sequence of values given by $(2j+1)^2 - 2(2j+1) - 4 = 4j^2 - 5$. And so on.

The first square-free product of two distinct primes from the set of primes congruent to +-1 modulo 5 is 209. For such a value of D we expect to find double the number of solutions for an individual prime, because solutions of the Pell equation are multiplicative based on the famous algebraic identity

$$(a^2 - 5b^2)(x^2 - 5y^2) = (ax + 5by)^2 - 5(ay+bx)^2$$

This enables us to "multiply" a solution for D=11 times a solution for D=19 to give a solution for D = (11)(19) = 209. Since there are two forward solutions for each prime, we can get four distinct product solutions. The Pell solutions for D=209 are

```
s[0] =   n  =  1   5   8  13  16  23  29  40  47   64   79  107...
s[1] =   m1 = 15  18  21  27  31  41  50  67  78  105  129  174 ...
```

These give the four distinct forward sequences for D=209

$$1 \quad 15 \quad 16 \quad 31 \quad 47 \quad 78 \ ...$$

$$5 \quad 8 \quad 23 \quad 41 \quad 64 \quad 105 \ ...$$

$$8 \quad 21 \quad 29 \quad 50 \quad 79 \quad 129 \ ...$$

$$13 \quad 27 \quad 40 \quad 67 \quad 107 \quad 174 \ ...$$

In general, if D is the product of k distinct primes congruent to +-1 (mod 5), ther there are 2^k distinct forward solution sequences, because each prime factor doubles the number of ways of multiplying the solutions. The reason multiplying by 1 or 5 doesn't double the solutions is that 1 and 5 each have just a single forward solution (because they are symmetrical).